Branchiura — a compendium of the geographical distribution and a summary of their biology

Authors' addresses:

L.A.M. Neethling, Department of Zoology, Faculty of Science, University of Johannesburg, P.O. Box 524, Auckland Park, Johannesburg, 2006, South Africa; e-mail: lneethling@uj.ac.za

A. Avenant-Oldewage, Department of Zoology, Faculty of Science, University of Johannesburg, P.O. Box 524, Auckland Park, Johannesburg, 2006, South Africa; e-mail: aoldewage@uj.ac.za

Manuscript first received 25 November 2015; final version accepted 6 September 2016.

Cover: Dolops ranarum (Stuhlmann, 1891), dorsal view of male; see p. 7, fig. 2D.

Branchiura — a compendium of the geographical distribution and a summary of their biology

By

Lourelle A.M. Neethling and Annemarè Avenant-Oldewage

CRUSTACEANA MONOGRAPHS, 21

BRILL

LEIDEN • BOSTON

The contents of this volume were originally published in 2016 in *Crustaceana* volume 89, issue 11-12.

This book is printed on acid-free paper.

ISBN13: 978 90 04 34615 4
E-ISBN: 978 90 04 34620 8

PRINTED IN THE NETHERLANDS

CONTENTS

> The page numbers in the above Table of Contents and Indices refer to the bracketed page numbers in this volume. The other page numbers are the page numbers in Crustaceana 89/11-12. When citing from this book, refer to Crustaceana 89 (2016) 1243-1446 and the page numbers without brackets.

[When citing this volume, please refer to Crustaceana 89 (2016) 1243-1446]

BRANCHIURA — A COMPENDIUM OF THE GEOGRAPHICAL DISTRIBUTION AND A SUMMARY OF THEIR BIOLOGY

BY

LOURELLE A. M. NEETHLING and ANNEMARIÈ AVENANT-OLDEWAGE

Department of Zoology, Faculty of Science, University of Johannesburg, P.O. Box 524, Auckland Park, Johannesburg, 2006, South Africa

ABSTRACT

This is a compendium of current knowledge about the crustacean subclass Branchiura Thorell, 1864. An overview of the group is presented, starting from the first species description, and reports of taxonomic changes. It also provides a condensed retrospect of each genus and includes the characteristics of each genus, the geographical distribution of each species arranged according to occurrence per continent; and aspects of the anatomy, physiology, host-parasite interactions and phylogeny are discussed. In order to condense the information available on members of the subclass, additional literature sources on each aspect are tabulated. The species lists provided by the World Register of Marine Species (WoRMS) were used as a starting point as these were the most comprehensive currently available.

RÉSUMÉ

Nous présentons ici une compilation des connaissances actuelles sur la sous-classe de Crustacés Branchiura Thorell, 1864. Un sommaire du groupe est exposé, en partant de la première description d'espèce et relate les divers changements taxonomiques. Ce travail fournit aussi un historique condensé de chaque genre, incluant les caractéristiques de chaque genre, la répartition géographique de chaque espèce traitée suivant la présence par continent ; les aspects de l'anatomie, de la physiologie, des interactions hôte-parasite et de la phylogénie sont discutés. Afin de condenser l'information disponible sur les membres de la sous-classe, des sources de littérature complémentaire sur chaque aspect sont fournies sous forme de tableaux. Les listes d'espèces fournies par le World Register of Marine Species (WoRMS) ont été utilisées comme point de départ, étant actuellement les plus exhaustives disponibles.

BRANCHIURA

A comprehensive review of the systematics of the Branchiura was published by Møller (2009), in which the author discussed the changes in taxonomic rank this group has undergone since its establishment. The current authors show how the group has evolved as the understanding of *Argulus*, the oldest described genus, changed; therefore, the systematics information that follows begins with the description of *Argulus*. It is important to note that the taxonomic ranks that follow are not the equivalent of the taxonomic ranks used today, and as such should not be used to compare. Furthermore, each new classification system is set out in the Appendix.

In Linnaeus' Systema Naturæ, tenth edition, published in 1758, the species assemblages known today as invertebrates were classified either in Class Insecta or the Class Vermes. Various crustacean species were grouped together in the Order Aptera (Class Insecta), with the genus *Monoculus* Linnaeus, 1758.

In response to this ambiguity, Müller (1785) established the section Entomostraca, an assemblage of Crustacea arranged according to the type of shell (bivalve or univalve), and the number of eyes (monoculus or binoculus) these animals possessed. Müller (1785) also erected the genus *Argulus* for species that have a single shell, a tail and setae on the legs.

Later, Latreille (1802) agreed with the system of classifying Crustacea based on the shell type, but further divided the Subclass Entomostraca into sections and orders. The genus *Argulus* was placed in the Order Pseudopoda because the head is fused with the thorax.

Thereafter, in a review of the systematics of Entomostraca, Leach (1819) rearranged the Entomostraca based on the structure of the appendages. Based on this system of classification, Leach (1819) erected the Family Argulidae under the Order Pæcillopoda due to the presence of a "beak-like mouth, four antennae and biramous appendages".

Following researchers before him, H. Milne Edwards (1834) agreed that the Crustacea should rather be classified according to the type of mouth they possess, and divided the Crustacea according to the presence or absence of special masticatory organs. This author placed the genus *Argulus* in the Order Siphonostoma because of its sucking mouthparts. Dana & Herrick (1837) disagreed with Milne Edwards's placement of *Argulus* because, as they explained, the siphon forms only a small part of the mouthparts of *Argulus* species. These authors suggested that argulids should rather be placed in an order between the Xiphosura and Siphonostoma, as the argulid mouthparts seemed to be a link between these two groups. Despite Dana and Herrick's argument, H. Milne Edwards (1840) maintained his previous classification.

Not long thereafter, in a review of the British Entomostraca, Baird (1850) followed a system similar to that of H. Milne Edwards (1834-1840), keeping the genus *Argulus* in the Order Siphonostoma.

Zenker (1854) later acknowledged the two systems of classifying Crustacea presented by Latreille and Milne Edwards. After going through each order individually and showing that the characteristics do not always match all the genera, he suggested that rather than a linear system, a circular system of classification should be employed. The author erected the Order Aspidostraca to place the genus *Argulus* and members of the order Branchiopoda Latreille (see Zenker, 1854).

Thereafter, Thorell (1864) reviewed the systematics of *Argulus* and agreed with the placement of the group within the Class Branchiopoda but, established the Order Branchiura with a single family, Argulidae Leach (see Thorell, 1864). Even though this addressed some of the inconsistencies with the classification of *Argulus*, some researchers such as Heller (1857) and Krøyer (1863) still chose to use the classification of H. Milne Edwards (1834, 1840).

In a further review of argulids, Leydig (1871) agreed with Thorell's system of placing the Branchiura as the third Order under the Class Branchiopoda along with Phyllopoda and Cladocera. Subsequently, after evaluating the larvae and adults, Claus (1875) placed the Branchiura as a third suborder under the Copepoda but retained the name Branchiura. Wilson (1902) followed this classification.

Later, Thiele (1904) suggested that the Branchiura should rather be classified as a separate group that is equal to the Copepoda and Branchiopoda. This suggestion was accepted by Grobben (1908) and later by Martin (1932) who further suggested raising the Branchiura to an independent subclass of the Crustacea.

Despite this, Yamaguti (1963) reclassified the members of the subclass Branchiura in his account of parasitic Copepoda and Branchiura of fishes. First, Yamaguti (1963) established a new order, Argulidea Yamaguti, 1963, whose diagnosis is a summary of the description given by Wilson (1932) for the suborder Arguloida. Within the Argulidea, Yamaguti (1963) placed two families; the first Argulidae Leach, 1819, and the second a new family, the Dipteropeltidae Yamaguti, 1963. The diagnosis of the Argulidae, remained as Leach (1819) had written, with the type genus being *Argulus* Müller, 1785. The family Argulidae was further divided into three new subfamilies. Argulinae Yamaguti, 1963 was diagnosed as Argulidae with two pairs of antennae, the presence of a preoral spine, "second maxilla" modified to form a prehensile disc and the basal plate of the "maxilliped" armed with "teeth" (Yamaguti, 1963), with *A. foliaceus* (Linnaeus, 1758) as the type species. Chonopeltinae Yamaguti, 1963, created for the genus *Chonopeltis* Thiele, 1900, was diagnosed as Argulidae with one pair of antennae, the absence of a preoral spine, a prehensile maxilla and the basal plate of the "maxilliped" without "teeth" (Yamaguti, 1963). And Dolopsinae Yamaguti, 1963 was diagnosed as Argulidae with two pairs of antennae, the "second maxilla" tipped with a claw and the presence or absence of the preoral spine (Yamaguti, 1963). This subfamily contained two genera namely *Dolops* Audouin, 1837 and *Huargulus* Yü, 1938. The second family, Dipteropeltidae was diagnosed as Argulidea with elongated lateral "wings" drawn from the carapace, the absence or rudimentary presence of the antennae and preoral spine, suckers without supporting rods and "maxillipeds" without basal plates (Yamaguti, 1963). This family contained two genera, *Dipteropeltis* Calman, 1912 and *Talaus* Moreira, 1913, each with only one species (Yamaguti, 1963). The genus *Talaus* was later synonymised with *Dipteropeltis* and the family is still valid to date.

Yamaguti's publication and classification of Branchiura and Copepoda drew attention because of his added complication of the classification, the use of genus characteristics for families, and misspelling of species names (Fryer, 1968, 1969; Møller, 2009). Publications that cited Yamaguti were often to correct mistakes he had made in the publication and his classification was never followed (Poly, 2008).

The most recent comprehensive classification was proposed by Martin and Davis (2001) and it was subsequently followed by Thatcher (2006), Boxshall (2007), Poly (2008) and Møller (2009)

Subclass Branchiura Thorell, 1864
  Order Arguloida Yamaguti, 1963
    Family Argulidae Leach, 1819
      Genus *Argulus* Müller, 1785
      Genus *Dolops* Audouin, 1837
      Genus *Chonopeltis* Thiele, 1900
      Genus *Dipteropeltis* Calman, 1912

Members of this subclass (fig. 1A-D) are ectoparasitic on marine and freshwater fish, tadpoles (Stuhlmann, 1891), salamanders (Paiva Carvalho, 1939; Poly, 2003) and alligators (Ringuelet, 1943, 1948) but are most commonly known as fish or carp lice (Piasecki & Avenant-Oldewage, 2008).

CHARACTERISTICS

Thorell (1864) described members of the Order Branchiura as follows: body is dorso-ventrally compressed; head forms a shield; compound eyes are separate; two pairs of antennae are short, first pair armed with a curved hook and shielded by the maxillipeds (sic); the mouth forms a siphon and may or may not include jaws; maxillipeds (sic) are strong and cup-like; four distinct thoracic segments, four pairs of biramous thoracopods; lack branchial appendages; the "tail" is unsegmented, smooth and fleshy and serves for respiration; metamorphosis is incomplete; external parasites of fish.

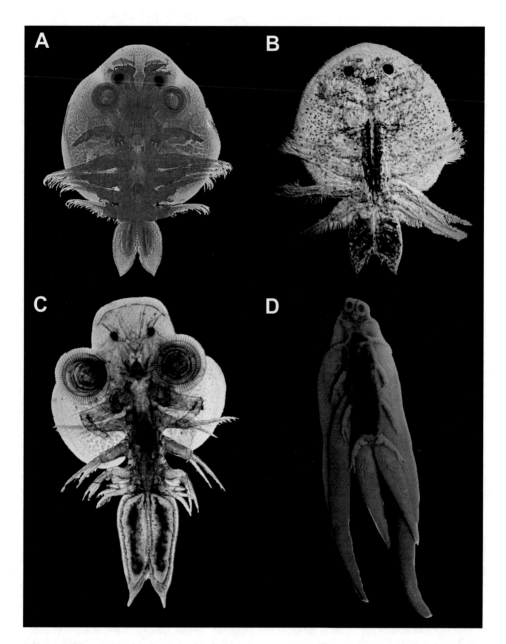

Fig. 1. Micrographs of representatives of Branchiura. A, ventral view of a male *Argulus coregoni* Thorell, 1864, stained with Lignin Pink; B, ventral view of male *Dolops ranarum* (Stuhlmann, 1891), micrograph taken by Dr. Quinton Tam; C, ventral view of male *Chonopeltis australis* Boxshall, 1976; D, ventral view of female *Dipteropeltis hirundo* Calman, 1912 from the British Museum of Natural History (BMNH 1892.10.24.2).

Much later, Martin (1932) updated the characteristics for what she then referred to as the Subclass Branchiura after studying *Argulus* spp.: the carapace forms a bilobed dorsal shield; the antennules and antennae are modified organs of attachment; the number of trunk somites is four and none of the somites are completely fused with the head; body unsegmented after genital aperture; body ends in caudal furcae; adult mandibles lack palps but enclosed in a suctorial proboscis; trunk limbs are biramous and the first two pairs carry a flabellum; they possess a pair of compound eyes and a nauplius eye; spermatozoa are transferred to the female spermathecae without the use of special copulatory organs or the formation of spermatophores.

The last characteristic was challenged by Fryer (1960b) after describing the spermatophore in *Dolops ranarum* (Stuhlmann, 1891), and again by Neethling & Avenant-Oldewage (2015) to read: spermatozoa are transferred to female spermathecae with a spermatophore, assisted by special copulatory structures on the legs of the males.

These characteristics however, do not apply to all four genera. In order for the characteristics to be inclusive of members of all four genera, the combined characteristics for Branchiura should read as: dorso-ventrally flattened bodies; cephalon and thorax fused and covered at least partly by a cephalic shield with carapace lobes that extend posteriorly (fig. 2A-D); abdomen fused and contains the testes or spermathecae. Appendages consist of a single pair of antennae (*Chonopeltis* spp., figs. 2B, 3C) or a pair of antennulae and a pair of antennae possibly modified for attachment when present (fig. 3B, D); one pair of separate compound eyes (fig. 2A, B, D) and a single nauplius eye; maxillulae modified to claws (*Dolops* sp., fig. 3A) or to suckers (figs. 2B, C, 3E); may possess a proboscis (*Argulus* spp., and *Dipteropeltis* spp.) with a retractible preoral spine anteriorly and the mouth posteriorly (figs. 2C, 3F); the mouth bears mandibles (fig. 3G) and may bear labial spines; maxillae (fig. 3H-K) usually prehensile, may have a basal plate

---

Fig. 2. Representatives of Branchiura to highlight morphological features. A, dorsal view of female *Argulus kosus* Avenant-Oldewage, 1994, 1 mm; B, ventral view of male *Chonopeltis victori* Avenant-Oldewage, 1991, 1 mm; C, ventral view of *Dipteropeltis campanaformis* Neethling, Malta & Avenant-Oldewage, 2014, 2 mm, copyright by Magnolia Press, reproduced here with permission from figures originally in Zootaxa 3755(2): 179-193; D, dorsal view of male *Dolops ranarum* (Stuhlmann, 1891). For all images, abbreviations are as follows: ab, abdomen; al, abdominal lobe; ans, antenna and antennule; ant, antenna; antl, antennule; ar, anterior respiratory area; ca, carapace; ce, compound eye; cs, cephalic shield; cl, carapace lobe; dr, dorsal ridge; fr, furcal rami; ir, interocular rods; mx, maxilla; mxl, maxillula; ne, nauplius eye; pr, posterior respiratory area; prb, proboscis; sp, spermatheca; sps, spermathecal spine; th, thorax; thp, thoracopod (function as swimming leg); ts, testis. Source: Avenant et al. (1989a), Avenant-Oldewage (1991, 1994a) (http://creativecommons.org/licenses/by/3.0/); Neethling et al. (2014).

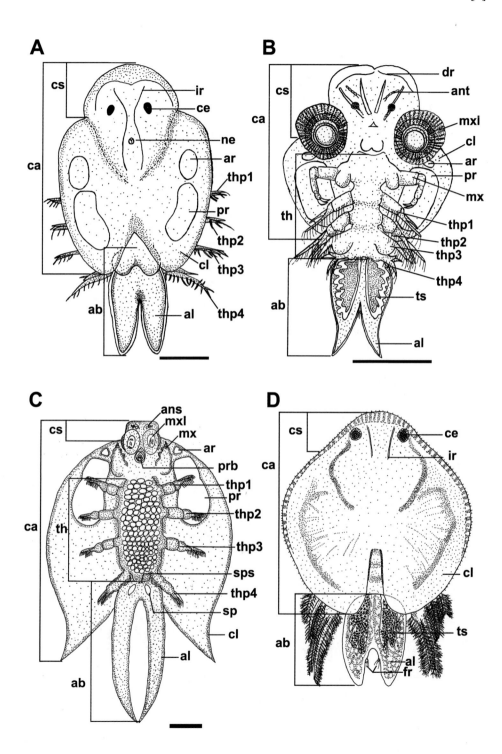

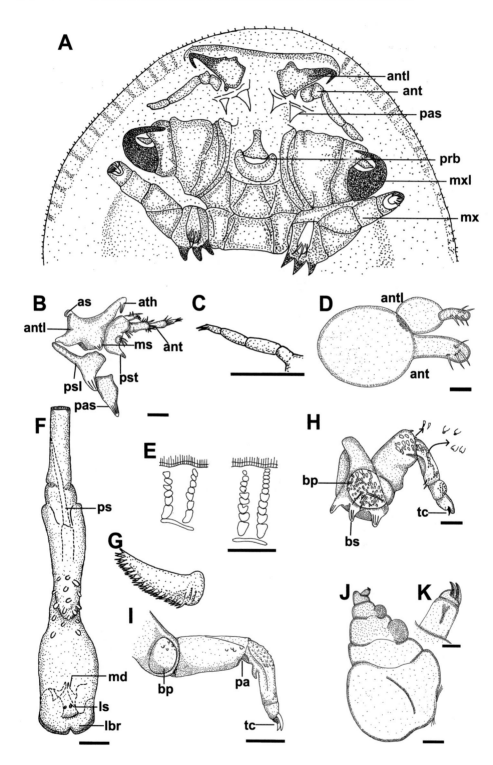

with basal spines (fig. 3H, I); four thoracic segments each with a pair of biramous thoracopods, the first two pairs may possess flabellum; in males the second, third (*Dolops* spp. fig. 4D) or fourth pairs of thoracopods, or a combination of these (*Argulus* spp., and *Chonopeltis* spp.), may possess unique accessory copulatory structures (fig. 4A-C); in females, the fourth pair of thoracopods carry a natatory lobe on the coxa (fig. 4E); the body terminates in a bilobed fused abdomen which may or may not possess furcal rami; exhibit sexual dimorphism; spermatozoa are transferred to female spermathecae (fig. 2A, C) with a spermatophore (observed in *Dolops, Argulus, Chonopeltis*), assisted by special copulatory structures on the thoracopods of the males when these structures are present. Respiratory areas present on the ventral surface of the carapace lobes in *Argulus, Chonopeltis* and *Dipteropeltis*.

THE GENERA OF THE SUBCLASS BRANCHIURA

Argulus Müller, 1785

O. F. Müller (1785) erected the genus *Argulus* Müller, 1785 and classified it as a univalve, binocular (two eyes) entomostracan crustacean with a tail and setae on the legs. The type species, *Argulus charon* Müller, 1785 was described from a juvenile specimen and the author provided a magnified sketch (Müller, 1785). Later, H. Milne Edwards (1840) listed *Argulus delphinus* Müller, 1785 as a synonym of *Monoculus foliaceus* Linnaeus, 1758 implying that this date and genus should take precedence; however, *Monoculus* has been suppressed by the International Commission on Zoological Nomenclature (ICZN) under the Opinion

Fig. 3. Individual anterior appendages of representatives of the Branchiura. A, anterior region of the female *Dolops ranarum* (Stuhlmann, 1891); B, antenna and antennule of *Argulus kosus* Avenant-Oldewage, 1994, 100 μm; C, antenna of *Chonopeltis victori* Avenant-Oldewage, 1991, 0.1 mm; D, antenna and antennule of *Dipteropeltis campanaformis* Neethling, Malta & Avenant-Oldewage, 2014, 20 μm, copyright by Magnolia Press, reproduced here with permission from figures originally in Zootaxa 3755(2): 179-193; E, sections of maxillula of *Argulus kosus* Avenant-Oldewage, 1994, 100 μm; F, proboscis of *Argulus kosus* Avenant-Oldewage, 1994, 100 μm; G, mandible of *Dolops ranarum* (Stuhlmann, 1891); H, maxilla of *Argulus kosus* Avenant-Oldewage, 1994, 100 μm; I, maxilla of *Chonopeltis victori* Avenant-Oldewage, 1991, 0.1 mm; J, maxilla of *Dipteropeltis campanaformis* Neethling, Malta & Avenant-Oldewage, 2014, 200 μm; K, terminal claw of maxilla of *Dipteropeltis campanaformis* Neethling, Malta & Avenant-Oldewage, 2014, 10 μm, copyright by Magnolia Press, reproduced here with permission from figures originally in Zootaxa 3755(2): 179-193. For all images, abbreviations are as follows: ant, antenna; antl, antennule; as, anterior spine; ath, antennule terminal hook; bp, basal plate; bs, basal spine; lbr, labrum; ls, labial spines; md, mandible; ms, medial spine; mx, maxilla; mxl, maxillula; pa, prehensile area (see Fryer, 1959; Van As, 1992); pas, post-antennular spine; prb, proboscis; ps, preoral spine; psl, posterior antennular spine; pst, posterior spine; tc, terminal claw. Source: Avenant et al. (1989a); Avenant-Oldewage (1991, 1994a) (http://creativecommons.org/licenses/by/3.0/); Neethling et al. (2014).

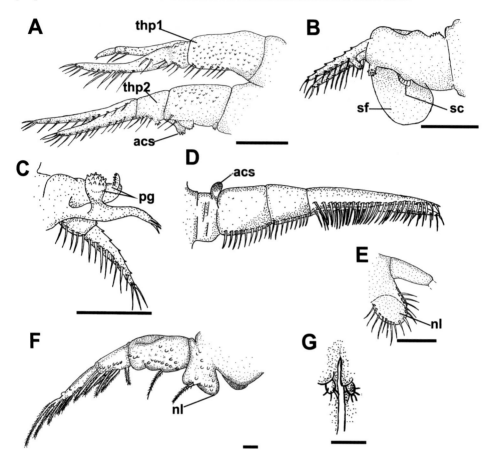

Fig. 4. Individual posterior appendages of representatives of the Branchiura. A, first and second thoracopod of male *Chonopeltis victori* Avenant-Oldewage, 1991, 0.1 mm; B, ventral view of the third thoracopod of *Chonopeltis victori* Avenant-Oldewage, 1991, 0.1 mm; C, dorsal view of the fourth thoracopod of the male *Chonopeltis victori* Avenant-Oldewage, 1991, 0.1 mm; D, male third thoracopod of *Dolops ranarum* (Stuhlmann, 1891); E, natatory lobe of the fourth thoracopod of female *Chonopeltis victori* Avenant-Oldewage, 1991, 0.1 mm; F, fourth thoracopod of female *Argulus kosus* Avenant-Oldewage, 1994, 100 μm; G, furcal rami of *Argulus kosus* Avenant-Oldewage, 1994, 100 μm. For all images, abbreviations are as follows: acs, accessory copulatory structures; nl, natatory lobe; pg, peg; sc, socket; sf, socket flap; thp, thoracopod. Sources: Avenant et al. (1989a); Avenant-Oldewage (1991, 1994a) (http://creativecommons.org/licenses/by/3.0/).

288 (ICZN, 1954) and is thereafter recognised as an unaccepted combination of *Argulus*.

Synonyms of the genus *Argulus* Müller, 1785

Monoculus Linnaeus, 1758
Binoculus Geoffroy Saint-Hilaire, 1762; *Agenor* Risso, 1827; *Argulus* see Thorell (1864).
Argulus Müller, 1785

Ozulus Latreille, 1802 (different order to *Argulus*); *Argulus* see Baird (1850) (review of Entomostraca).

Diprosia Rafinesque-Schmaltz, 1814; *Argulus* see Holthuis (1954).

Agenor Risso, 1827; *Argulus* see Thorell (1864).

Huargulus Yü, 1938; *Argulus* see Ku & Wang (1956) and Fryer (1969) (based on a juvenile *Argulus* specimen).

Synonyms of the species of *Argulus* Müller, 1785
Argulus ambloplites jollymani (Fryer, 1960a)

Argulus jollymani Fryer, 1956
Argulus ambloplites jollymani (Fryer, 1960a)

Argulus appendiculosus Wilson, 1927

Argulus appendiculosus Wilson, 1907. Wilson (1916); Pearse (1924); Tidd (1931); Schumacher (1952); Benda (1969); Cressey (1972, 1978); Dechtiar (1972); Poly (1998a).
Argulus biramosus Bere, 1931 according to Meehean (1940). Allum & Hugghins (1959); Dechtiar & Nepzy (1988).

Argulus arcassonensis Cuénot, 1912

Argulus arcassonensis Cuénot, 1912
Argulus otolithi Brian, 1927 according to Rushton-Mellor (1994b).
Argulus zei Brian, 1924 synonym according to Monod (1928).

Argulus flavescens Wilson, 1916

Argulus flavescens Wilson, 1916. Mueller (1936); Bangham (1940); Roberts (1957); Cressey (1972); Poly (1997, 1998a).
Argulus piperatus Wilson, 1920 according to Meehean (1940).

Argulus foliaceus (Linnaeus, 1758)

Argulus foliaceus (Linnaeus, 1758). Zaddach (1844); Scott & Scott (1913); Herter (1926); Chen (1933); Bower-Shore (1940); Wang (1958); Bazal et al. (1969); Mishra & Chubb (1969); Rizvi (1969); Czeczuga (1971); Gattaponi (1971); Grabda (1971); Van den Bosch de Aguilar (1972); Wingstrand (1972); Laurent (1975); Canić et al. (1977); Moravec (1978); Schlüter (1979); Hallberg (1982); Zhiliukas & Rauckis (1982); Økland (1985); Ali et al. (1988); Køie (1988); Khalifa (1989); Menezes et al. (1990); Kubrakiewicz & Klimowicz (1994); Buchmann et al. (1995); Buchmann & Bresciani (1997); Northcott et al. (1997); Mikheev et al. (1998, 2000); Molnár & Szèkely (1998); Pasternak et al. (2000); Cengizler et al. (2001); Gault et al. (2002); Yildiz & Kumantas (2002); Thilakaratne et al. (2003); Özan & Kir (2005); Öztürk (2005); Tekin Özan & Kir (2005); Harrison et al. (2006, 2007); Karatoy & Soylu (2006); Öktener et al. (2006, 2007); Uzunay & Soylu (2006); Jalali & Barzegar (2006); Møller et al. (2007, 2008); Jalali et al. (2008); Møller (2009); Taylor et al. (2009a, b); Alaş et al. (2010); Noaman et al. (2010); Macchioni et al. (2015).
Argulus rothschildi Leigh-Sharpe, 1933 according to Gurney (1948).
Argulus viridis Nettovich, 1900 according to Romanovsky (1955).

Argulus funduli Krøyer, 1863

Argulus funduli Krøyer, 1863. Wilson (1902, 1907, 1932); Bere (1936); Cressey (1972, 1978); Poulin (1999).
Argulus latus Smith, 1873 according to Meehean (1940). Wilson (1902, 1932, 1944).

Argulus japonicus Thiele, 1900

Argulus japonicus Thiele, 1900. Tokioka (1936b); Yamaguti (1937); Tokioka (1939); Wilson (1944); Hsiao (1950); Fryer (1960a); Heegard (1962); Pilgrim (1967); Bazal et al. (1969); Grabda (1971); Cressey (1972, 1978); Amin (1981); Kruger et al. (1983); Shimura (1983b); Seng (1986); Nagasawa et al. (1989); Jafri & Ahmed (1991); LaMarre & Cochran (1992); Gresty et al. (1993); Bunkley-Williams & Williams (1994); Ikuta & Makioka (1994, 1997); Lutsch & Avenant-Oldewage (1995); Ikuta et al. (1997); Han et al. (1998); Avenant-Oldewage (2001); Tam & Avenant-Oldewage (2006, 2009a); Wadeh et al. (2008); Yoshizawa & Nogami (2008); Sahoo et al. (2012); Alsarakibi et al. (2014).
Argulus matritensis Arévalo, 1921 accordings to Fryer (1982).
Argulus pellucidus Wagler, 1935 according to Fryer (1960a); Hoffman (1967).

Argulus kosus Avenant-Oldewage, 1994

Argulus kosus Avenant-Oldewage, 1994. Van As et al. (1999).
Argulus smalei Avenant-Oldewage & Oldewage, 1995 according to Van As et al. (1999).

Argulus megalops Smith, 1873

Argulus megalops Smith, 1873. Wilson (1902, 1932); Causey (1960); Davis (1965); Cressey (1972, 1978).
Argulus megalops var. *spinosus* Wilson, 1944 according to Meehean (1940).
Argulus varians Bere, 1936 according to Meehean (1940). Bouchet (1985).

Argulus nattereri Heller, 1857

Argulus nattereri Heller, 1857. Krøyer (1863); Wilson (1902); Moreira (1912, 1913); Ringuelet (1943, 1948); Brian (1947); Mamani et al. (2004).
Argulus silvestrii Lahille, 1926 according to Ringuelet (1943, 1948).

Argulus nobilis Thiele, 1904

Argulus nobilis Thiele, 1904. Wilson (1924); Cressey (1972).
Argulus ingens Wilson, 1912 as stated by Wilson (1924); Meehean (1940).

Argulus pugettensis Dana, 1853

Argulus pugettensis Dana, 1853. Wilson (1902, 1909, 1924); Yamaguti (1963); Cressey (1972).
Argulus niger Wilson, 1902 according to Meehean (1940). Wilson (1909).

Argulus purpureus (Risso, 1826)

Argulus purpureus (Risso, 1826)
Argulus purpureus (Risso, 1816), see Poly (1998b).
Argulus giganteus Lucas, 1849 according to Wilson (1902) and Rushton-Mellor (1994b).

Argulus salmini Krøyer, 1863

Argulus salmini Krøyer, 1863. Wilson (1902); Moreira (1912, 1913); Ringuelet (1943, 1948).
Argulus paulensis Wilson, 1924 according to Meehean (1940).

Argulus stizostethii Kellicott, 1880

Argulus stizostethii Kellicott, 1880. Wilson (1916); Poly (1998a).
Argulus canadensis Wilson, 1916 according to Meehean (1940). Sandeman & Pippy (1967); Watson & Dick, 1979; Poulin & Fitzgerald (1987, 1988a, b, 1989a, b, c); Dugatkin et al. (1994).

Argulus vittatus (Rafinesque-Schmaltz, 1814)

Argulus vittatus (Rafinesque-Schmaltz, 1814)
Argulus purpureus (Risso, 1826) according to Thorell (1864), Holthuis (1954) and Poly (1998b).
Argulus giganteus Lucas, 1849 according to Boxshall & Walter (2009) and Ramdane & Trilles (2012).

Argulus species and their distribution

The genus *Argulus* is the most speciose genus in the subclass Branchiura (see tables I–VI). In a review of the *Argulus* species housed at the United States National Museum, Meehean (1940) cautioned against overzealous differentiating between species, having found that certain species were actually synonyms or variations of others (see tables I–VI); evidence of this was found in literature when compiling the distribution tables. According to WoRMS there are 159 *Argulus* species with 17 synonymised species (Boxshall, 2009). The species list provided by WoRMS was used as a starting point from which to summarize the distribution; additional information was obtained from literature and is presented in the following tables according to the continent on which they were recorded or found.

Research conducted on *Argulus* species

Research on the biology of *Argulus* spp. encompasses a variety of aspects that can be grouped under three topics that include the anatomy and physiology of the species (fig. 5A-C), host-parasite interactions (fig. 5D-E), and phylogeny.

Anatomy and physiology of *Argulus* species

Table VII lists a selection of references available to the authors of publications on *Argulus* species.

TABLE I

Marine, brackish and freshwater *Argulus* spp. recorded from the continent of North America

Argulus species	Country/city/town	Water system as published	Water system confirmed[a]	E	Host species	Host species confirmed[b]	Reference
Argulus alosae Gould, 1841	Massachusetts, U.S.A.	Unknown	Unknown	M	*Alosa vulgaris* Alewife	*Alosa pseudoharengus* (Wilson, 1811)	Gould (1841)
	Unknown	Unknown	Unknown	M	*Pomolobus pseudoharengus* Wilson	*Alosa pseudoharengus* (Wilson, 1811)	Wilson (1902)
	Massachusetts, U.S.A.	Woods Hole	Woods Hole, North Atlantic Ocean	M	Smelt	*Osmerus eperlanus* (Linnaeus, 1758)	
	New Jersey, U.S.A.	Great Egg Harbour	Great Egg Harbour, North Atlantic Ocean	M	Not given	Not given	
	Key West, Florida, U.S.A.	Key West	Gulf of Mexico, Atlantic Ocean	M	Not given	Not given	
	Massachusetts, U.S.A.	Woods Hole	Woods Hole, North Atlantic Ocean	M	Not given	Not given	
	Long Island, New York, U.S.A.	Patchogue River	Patchogue River	B	Not given	Not given	
	Quebec, Canada	Gulf of St. Lawrence	Gulf of Saint Lawrence, Atlantic Ocean	B	*Gasterosteus biaculeatus* Shaw	*Gasterosteus aculeatus* Linnaeus, 1758	
	Maine, New England, U.S.A.	Casco Bay	Casco Bay, Gulf of Maine, North Atlantic Ocean	M	*Ctenolabrus adspersus* Walbaum "common cunner"	*Tautogolabrus adspersus* (Walbaum, 1792)	Wilson (1904)
	Nova Scotia, Canada	Shubenacadie River	Shubenacadie River, Minas Basin	B	Free swimming	Free swimming	Wilson (1920a)

TABLE I
(Continued)

Argulus species	Country/city/town	Water system as published	Water system confirmed[a]	E	Host species	Host species confirmed[b]	Reference
	Nova Scotia, Canada	Bass River	Bass River, Minas Basin, Nova Scotia	F	Microgadus tomcod	Microgadus tomcod (Walbaum, 1792)	Wilson (1924)
	Massachusetts, U.S.A.	Woods Hole	Woods Hole, North Atlantic Ocean	M	Pomolobus pseudoharengus	Alosa pseudoharengus (Wilson, 1811)	Wilson (1932)
	Massachusetts, U.S.A.	Woods Hole	Woods Hole, North Atlantic Ocean	M	Osmerus mordax	Osmerus mordax mordax (Mitchill, 1814)	
	Maine to Texas	East coast of the U.S.A.	North Atlantic Ocean	M	Alosa	Alosa sp.	Cressey (1972)
				M	Clupea	Clupea sp.	
				M	Dorosoma	Dorosoma sp.	
				M	Tautogolabrus	Tautogolabrus sp.	
				M	Lepisosteus	Lepisosteus sp.	
				M	Opsanus	Opsanus sp.	
				M	Cynoscion	Cynoscion sp.	
	Old Saybrook, Connecticut, U.S.A.	Connecticut River	Connecticut River	B	Brevoortia tyrannus (Latrobe) Atlantic Menhaden	Brevoortia tyrannus (Latrobe, 1802)	Kroger & Guthrie (1972)
	Nova Scotia to the Gulf coast of Texas	East coast of the U.S.A.	North Atlantic Ocean	M	Microgadus tomcod	Microgadus tomcod (Walbaum, 1792)	Cressey (1978)
				M	Tautogolabrus adspersus	Tautogolabrus adspersus (Walbaum, 1792)	
				M	Alosa pseudoharengus	Alosa pseudoharengus (Wilson, 1811)	

TABLE I
(Continued)

Argulus species	Country/city/town	Water system as published	Water system confirmed[a]	E	Host species	Host species confirmed[b]	Reference
				M	Dorosoma cepedianum	Dorosoma cepedianum (Lesueur, 1818)	
				M	Clupea harengus	Clupea harengus Linnaeus, 1758	
				M	Cynoscion nebulosus	Cynoscion nebulosus (Cuvier, 1830)	
				M	Gasterosteus sp.	Gasterosteus sp.	
				M	Strongylura marina	Strongylura marina (Walbaum, 1792)	
				M	Rhopilema verrilli	Rhopilema verrilli (Fewkes, 1887) mushroom jellyfish	
Argulus americanus Wilson, 1902	Ann Arbor, Michigan		Huron River	F	Amia calva Linnaeus	Amia calva Linnaeus, 1766	Wilson (1902)
	Culver, Indiana, Marshall County, U.S.A.	Lake Maxinkuckee	Lake Maxinkuckee	F	Amia calva Linnaeus	Amia calva Linnaeus, 1766	Wilson (1916)
	Fairport, Muscatine County, Iowa, U.S.A.	Fairport	Mississippi River	B	Amia calva Linnaeus	Amia calva Linnaeus, 1766	
				B	Umbra limi	Umbra limi (Kirtland, 1840)	
	Tennessee, U.S.A.	Reelfoot Lake	Reelfoot Lake	F	Amia calva Linnaeus Grindle, Bowfin	Amia calva Linnaeus, 1766	Bangham & Venard (1942)
				F	Ameiurus nebulosus (LeSueur)	Ameiurus nebulosus (Lesueur, 1819)	

TABLE I
(Continued)

Argulus species	Country/city/town	Water system as published	Water system confirmed[a]	E	Host species	Host species confirmed[b]	Reference
				F	*Sclerotis miniatus* (Jordan)	*Lepomis miniatus* (Jordan, 1877)	
				F	*Helioperca macrochira* (Rafinesque)	*Lepomis macrochirus* Rafinesque, 1819	Goin & Ogren (1956)
	Gainesville, Florida, U.S.A.	Not given	Not given	F	*Pseudobranchus s. axanthus* Perennibranch salamander	*Pseudobranchus axanthus* (Netting & Goin, 1942) Narrow-striped Dwarf Siren	Goin & Ogren (1956)
	Gainesville, Florida	Newmans Lake	Newmans Lake	F	*Rana heckscheri* Wright (frog)	*Lithobates heckscheri* (Wright, 1924) River frog	
	Wisconsin, U.S.A.	Unknown	Unknown	F	*Amia calva* Bowfin	*Amia calva* Linnaeus, 1766	Shimura & Asai (1984)
	Williams County, Illinois, U.S.A.	Lake Creek	Lake Creek	F	*Amia calva* Linnaeus, 1766	*Amia calva* Linnaeus, 1766	Poly (1998a)
	Williams County, Illinois, U.S.A.	Little Grassy Creek	Little Grassy Creek	F	*Amia calva* Linnaeus, 1766	*Amia calva* Linnaeus, 1766	
	Union County, Illinois, U.S.A.	LaRue Marsh	La Rue Swamp	F	*Amia calva* Linnaeus, 1766	*Amia calva* Linnaeus, 1766	
Argulus ambystoma Poly, 2003	Michoacán, Mexico	Lake Pátzcuaro	Lake Pátzcuaro	B	*Ambystoma dumerilii* (Dugès)	*Ambystoma dumerilii* (Dugès, 1870) Lake Patzcuaro Salamander	Poly (2003)

TABLE I
(Continued)

Argulus species	Country/city/town	Water system as published	Water system confirmed[a]	E	Host species	Host species confirmed[b]	Reference
Argulus appendiculosus Wilson, 1907	Montpelier, Vermont, U.S.A.	Unknown	Winooski River	F	sucker	Catostomus commersonii (Lacépède, 1803)	Wilson (1907)
	Kentucky	Cumberland Falls	Cumberland Falls, Cumberland River	F	Ictalurus punctatus	Ictalurus punctatus (Rafinesque, 1818)	Wilson (1916)
		Cumberland River	Cumberland River	F	A. grunniens Sheepshead	Aplodinotus grunniens Rafinesque, 1819	
	Fairport, Muscatine County, Iowa, U.S.A.	Fairport	Mississippi River	B	M. salmoides largemouth black bass	Micropterus salmoides (Lacépède, 1802)	
				F	Ictiobus cyprinella redmouth buffalo	Ictiobus cyprinellus (Valenciennes, 1844)	
				B	Ictiobus bubalus smallmouth buffalo	Ictiobus bubalus (Rafinesque, 1818)	
	Not given	Not given	Not given	M	Dorosoma cepedianum gizzard shad	Dorosoma cepedianum (Lesueur, 1818)	
				F	Pomoxis annularis crappie	Pomoxis annularis Rafinesque, 1818	
				F	Roccus chrysops white bass	Morone chrysops (Rafinesque, 1820)	
	Wisconsin, U.S.A.	Lake Pepin	Lake Pepin	F	mudcat	Ameiurus nebulosus (Lesueur, 1819)	Pearse (1924)

TABLE I
(Continued)

Argulus species	Country/city/town	Water system as published	Water system confirmed[a]	E	Host species	Host species confirmed[b]	Reference
	Ohio, U.S.A.	Lake Erie	Lake Erie	F	Ameiurus nebulosus Bullhead	Ameiurus nebulosus (Lesueur, 1819)	Tidd (1931)
	Jackson County, Minnesota, U.S.A.	Clear Lake	Clear Lake	F	Stizostedion vitreum vitreum	Sander vitreus (Mitchill, 1818)	Schumacher (1952)
				F	Ameiurus sp. bullheads	Ameiurus sp.	
	Wright County, Minnesota, U.S.A.	Collingwood Lake	Collingwood Lake	F	Stizostedion vitreum vitreum	Sander vitreus (Mitchill, 1818)	
	Pike County, Petersburg, Indiana	White River	Lower White River	F	Lepisosteus osseus Longnose gar	Lepisosteus osseus (Linnaeus, 1758)	Benda (1969)
	Ontario, Canada	Lake Erie	Lake Erie	F	Cyprinus carpio	Cyprinus carpio Linnaeus, 1758	Dechtiar (1972)
	Vermont, Michigan, Kentucky, Iowa, Wisconsin, Texas, Wyoming, South Dakota	Not given	Not given	F	Stizostedion	Sander sp.	Cressey (1972)
				F	Ictalurus	Ictalurus sp.	
				F	Micropterus	Micropterus sp.	
				F	Ictiobus	Ictiobus sp.	
				F	Catostomus	Catostomus sp.	
				F	Dorosoma	Dorosoma sp.	
	Vermont to Virginia west to Wyoming and also Texas and Louisiana	Not given	Not given	B	Catostomus catostomus	Catostomus catostomus (Forster, 1773)	Cressey (1978)
				B	Ictiobus cyprinellus	Ictiobus cyprinellus (Valenciennes, 1844)	
				B	Ictiobus bubalis	Ictiobus bubalus (Rafinesque, 1818)	

TABLE I
(Continued)

Argulus species	Country/city/town	Water system as published	Water system confirmed[a]	E	Host species	Host species confirmed[b]	Reference
				B	Ictalurus punctatus	Ictalurus punctatus (Rafinesque, 1818)	
				B	Ictalurus melas	Ameiurus melas (Rafinesque, 1820)	
				B	Ictalurus nebulosus	Ameiurus nebulosus (Lesueur, 1819)	
				B	Amia calva	Amia calva Linnaeus, 1766	
				B	Micropterus salmoides	Micropterus salmoides (Lacépède, 1802)	
				B	Morone (Roccus) chrysops	Morone chrysops (Rafinesque, 1820)	
				B	Perca flavescens	Perca flavescens (Mitchill, 1814)	
				B	Cyprinus carpio	Cyprinus carpio Linnaeus, 1758	
				B	Stizostedion vitreum	Sander vitreus (Mitchill, 1818)	
				B	Dorosoma cepedianum	Dorosoma cepedianum (Lesueur, 1818)	
				B	Lepisosteus osseus	Lepisosteus osseus (Linnaeus, 1758)	
	Franklin County, Illinois, U.S.A.	Rend Lake	Rend Lake	F	Free swimming	Free swimming	Buttner (1980)

TABLE I
(Continued)

Argulus species	Country/city/town	Water system as published	Water system confirmed[a]	E	Host species	Host species confirmed[b]	Reference
	Racine County, Southeast Wisconsin, U.S.A.	Tichigan Lake	Tichigan Lake	F	*Ictalurus punctatus* (Raf., channels catfish)	*Ictalurus punctatus* (Rafinesque, 1818)	Amin (1981)
		Fox River, Near Tichigan Lake	Fox River	F	*Roccus chrysops* (Raf., white bass)	*Morone chrysops* (Rafinesque, 1820)	
	Iowa, U.S.A.	Little Sioux River	Little Sioux River	F	*Cyprinus carpio*	*Cyprinus carpio* Linnaeus, 1758	Sutherland & Wittrock (1986)
	Between Falls and Berwick, Pennsylvania, U.S.A.	Susquehanna River	Susquehanna River	F	*Esox masquinongy* muskellunge	*Esox masquinongy* Mitchill, 1824	Deutsch (1989)
	Between Mocanaqua and Bloomsburg, Pennsylvania, U.S.A.	Susquehanna River	Susquehanna River	F	*Esox masquinongy* muskellunge	*Esox masquinongy* Mitchill, 1824	
	Dauphin, Manitoba, Canada	Dauphin Lake (51°17'N 99°48'W)	Dauphin Lake	F	*Stizostedion vitreum* Mitchell	*Sander vitreus* (Mitchill, 1818)	Szalai & Dick (1991)
				F	*Esox lucius* Linnaeus	*Esox lucius* Linnaeus, 1758	
				F	*Catostomus commersoni* Lacépède	*Catostomus commersonii* (Lacépède, 1803)	
				F	*Carpiodes cyprinus* LeSueur	*Carpiodes cyprinus* (Lesueur, 1817)	
				F	*Coregonus artedii* LeSueur	*Coregonus artedi* Lesueur, 1818	

TABLE I
(Continued)

Argulus species	Country/city/town	Water system as published	Water system confirmed[a]	E	Host species	Host species confirmed[b]	Reference
	Ohio, U.S.A.			F	Netropis atherinoides Rafinesque	Netropis atherinoides Rafinesque, 1818	Poly (1997)
		Muskingum River	Muskingum River	F	Pylodictis olivaris	Pylodictis olivaris (Rafinesque, 1818)	
		Great Miami River	Great Miami River	F	Micropterus dolomieu	Micropterus dolomieu Lacépède, 1802	
				F	Ictalurus punctatus	Ictalurus punctatus (Rafinesque, 1818)	
				F	Cyprinus carpio	Cyprinus carpio Linnaeus, 1758	
		Buck Creek	Buck Creek	F	Cyprinus carpio	Cyprinus carpio Linnaeus, 1758	
		Salt Creek	Salt Creek	F	Noturus miurus	Noturus miurus Jordan, 1877	
		Sunfish Creek	Sunfish Creek	F	Ictalurus punctatus	Ictalurus punctatus (Rafinesque, 1818)	
	Williams County, Illinois, U.S.A.	Becks Creek	Becks Creek	F	Pylodictis olivaris	Pylodictis olivaris (Rafinesque, 1818)	Poly (1998a)
		Cache River	Cache River	F	Host unknown	Host unknown	
Argulus bicolor Bere, 1936	Gulf of Mexico, Florida	Lemon Bay	Gulf of Mexico, Atlantic Ocean	M	Strongylura notata needlefish	Strongylura notata notata (Poey, 1860)	Bere (1936)
				M	Archosargus unimaculatus sand bream	Archosargus rhomboidalis (Linnaeus, 1758)	
	North Carolina to Louisiana	Southeast coast of the U.S.A.	North Atlantic Ocean	M	Strongylura	Strongylura sp.	Cressey (1972)

TABLE I
(Continued)

Argulus species	Country/city/town	Water system as published	Water system confirmed[a]	E	Host species	Host species confirmed[b]	Reference
				M	Morone	Morone sp.	
				M	Gobionellus	Gobionellus sp.	
				M	Micropogon	Micropogon sp.	
				M	Scamboromorus	Scamboromorus sp.	
				M	Dorosoma	Dorosoma sp.	
				M	Rhinoptera	Rhinoptera sp.	
Argulus biramosus Bere, 1931[c]	Vilas County, Wisconsin, U.S.A.	Little Star Lake	Little Star Lake	F	yellow perch	Perca flavescens (Mitchill, 1814)	Bere (1931)
	Lake County, South Dakota, U.S.A.	Brandt Lake	Brandt Lake	F	Catostomus commersoni white sucker	Catostomus commersonii (Lacépède, 1803)	Allum & Hugghins (1959)
				F	Cyprinus carpio carp	Cyprinus carpio Linnaeus, 1758	
				F	Ictalurus melas black bullhead	Ameiurus melas (Rafinesque, 1820)	
				F	Perca flavescens yellow perch	Perca flavescens (Mitchill, 1814)	
				F	Stizostedion v vitreum walleye	Sander vitreus (Mitchill, 1818)	
	Hamlin County, South Dakota, U.S.A.	Lake Poinsett	Lake Poinsett	F	Ictiobus cyprinella largemouth buffalo	Ictiobus cyprinellus (Valenciennes, 1844)	
	Ontario, Canada	Lake Erie	Lake Erie	F	Ictalurus punctatus (Rafinesque) channel catfish	Ictalurus punctatus (Rafinesque, 1818)	Dechtiar & Nepzy (1988)
Argulus borealis Wilson, 1912	Nainamo, British Columbia		Pacific Ocean	M	Lepidopsetta bilineata	Lepidopsetta bilineata (Ayres, 1855)	Wilson (1912a)

TABLE I
(Continued)

Argulus species	Country/city/town	Water system as published	Water system confirmed[a]	E	Host species	Host species confirmed[b]	Reference
	Northwest coast of the U.S.A.	Northwest coast of the U.S.A.	Pacific Ocean	M	*Lepidopsetta*	*Lepidopsetta* sp.	Cressey (1972)
Argulus canadensis Wilson, 1916[d]	Le Claire, Lake of the Woods County, Minnesota, U.S.A.	Lake of the Woods	Lake of the Woods	M	*Cymatogaster*	*Cymatogaster* sp.	Wilson (1916)
				F	*Coregonus* sp.	*Coregonus* sp.	
				B	*Acipenser rubicundus* rock sturgeon	*Acipenser sturio* Linnaeus, 1758	
	Newfoundland, Canada	Ocean Pond	Ocean Pond	F	*Salvelinus fontinalis*	*Salvelinus fontinalis* (Mitchill, 1814)	Sandeman & Pippy (1967)
		Small gully south of Ocean Pond	Small gully south of Ocean Pond	F	*Salvelinus fontinalis*	*Salvelinus fontinalis* (Mitchill, 1814)	
		Ocean Pond	Ocean Pond	F	*Salmo solar*	*Salmo salar* Linnaeus, 1758	
	Thompson, Manitoba, Canada	Southern Indian Lake	Southern Indian Lake	B	*Coregonus clupeaformis* "whitefish" (Mitchell)	*Coregonus clupeaformis* (Mitchill, 1818)	Watson & Dick (1979)
				B	*Coregonus arteddii* "cisco" Lesueur	*Coregonus artedi* Lesueur, 1818	
	Isle Verte, Quebec	salt marsh (48°01′N 69°21′W)	salt marsh (48°01′N 69°21′W)	B	*Gasterosteus aculeatus*	*Gasterosteus aculeatus* Linnaeus, 1758	Poulin & Fitzgerald (1987, 1988a, b; 1989a, b, c)

TABLE I
(Continued)

Argulus species	Country/city/town	Water system as published	Water system confirmed[a]	E	Host species	Host species confirmed[b]	Reference
	Isle Verte, Quebec	salt marsh (48°01′N 69°21′W)	salt marsh (48°01′N 69°21′W)	B	*Gasterosteus wheatlandi*	*Gasterosteus wheatlandi* Putnam, 1867	Poulin & Fitzgerald (1987, 1988a, b, 1989b, c)
	Isle Verte, Quebec	salt marsh (48°01′N 69°21′W)	salt marsh (48°01′N 69°21′W)	B	*Pungitius pungitius*	*Pungitius pungitius* (Linnaeus, 1758)	Poulin & Fitzgerald (1987, 1989b)
	Isle Verte, Quebec	tidal pool	tidal pool	B	*Gasterosteus aculeatus*	*Gasterosteus aculeatus* Linnaeus, 1758	Dugatkin et al. (1994)
Argulus catastomi Dana & Herrick, 1837	Whitneyville, New Haven	Mill River	Mill River	F	*Catastomus* sp.	*Catostomus* sp.	Dana & Herrick (1837)
	New York	Cayuga Lake	Cayuga Lake	F	suckers	*Catostomus commersonii* (Lacépède, 1803)	Kellicott (1886)
	Warren, Worcester County, Massachusetts, U.S.A.		Quaboag River	F	*Catastomus bostonensis* LeSeur	*Catostomus commersonii* (Lacépède, 1803)	Wilson (1902)
	Chicopee, Hampden County, Massachusetts, U.S.A.		Connecticut River	F	*Catastomus bostonensis* LeSeur	*Catostomus commersonii* (Lacépède, 1803)	
	Fairbury, Livingston County, Illinois, U.S.A.		Indian Creek	F	Carp	*Cyprinus carpio* Linnaeus, 1758	
				F	*Erimyzon sucetta oblongus* Mitchell	*Erimyzon sucetta* (Lacépède, 1803)	

TABLE I
(Continued)

Argulus species	Country/city/town	Water system as published	Water system confirmed[a]	E	Host species	Host species confirmed[b]	Reference
	Swanton, Vermont	Missequoi River	Missisquoi River	F	*Catastomus catastomus*	*Catostomus catostomus* (Forster, 1773)	Wilson (1907, 1916)
				F	*Catastomus nigricans*	*Hypentelium nigricans* (Lesueur, 1817)	
	Culver, Indiana, Marshall County, U.S.A.	Lake Maxinkuckee	Lake Maxinkuckee	F	*Catastomus nigricans*	*Hypentelium nigricans* (Lesueur, 1817)	
	Walworth County, Wisconsin, U.S.A.	Lake Geneva	Geneva Lake	F	suckers	*Catostomus commersonii* (Lacépède, 1803)	Pearse (1924)
	Madison, Wisconsin, U.S.A.	Lake Mendota	Lake Mendota	F		*Catostomus commersonii* (Lacépède, 1803)	
	New Haven, Connecticut	Long Island Sound	Long Island Sound	B	*Catastomus commersonii*	*Catostomus commersonii* (Lacépède, 1803)	Wilson (1932)
	Massachusetts, U.S.A.	Woods Hole	Woods Hole, North Atlantic Ocean	M	*Catastomus commersonii*	*Catostomus commersonii* (Lacépède, 1803)	
	Wisconsin, U.S.A.	Muscallonge Lake	Muskallonge Lake	F	*Catastomus commersonnii* (Lacépède) Common white sucker	*Catostomus commersonii* (Lacépède, 1803)	Bangham (1946)

TABLE I
(Continued)

Argulus species	Country/city/town	Water system as published	Water system confirmed[a]	E	Host species	Host species confirmed[b]	Reference
	Sawyer County, Wisconsin, U.S.A.	Lost Land Lake	Lost Land Lake	F	Catastomus c. commersonnii (Lacépède) Common white sucker	Catostomus commersonii (Lacépède, 1803)	Fischthal (1947)
	Washburn County, Wisconsin, U.S.A.	Birch Lake	Birch Lake	F	Catastomus c. commersonnii (Lacépède) Common white sucker	Catostomus commersonii (Lacépède, 1803)	Fischthal (1950)
				F	Perca flavescens (Mitchell) Yellow Perch	Perca flavescens (Mitchill, 1814)	
	Northeast U.S. from Minnesota to Vermont and south as far as Maryland	Not given	Not given	F	Catastomus	Catostomus sp.	Cressey (1972)
				F	cyprinids	Cyprinid	
	Northern U.S. from Massachusetts west to Minnesota	Not given	Not given	F	Catostomus commersonii	Catostomus commersonii (Lacépède, 1803)	Cressey (1978)
				F	Catostomus catostomus	Catostomus catostomus (Forster, 1773)	
				F	Hypentelium nigricans	Hypentelium nigricans (Lesueur, 1817)	
				F	Erimyzon sucetta	Erimyzon sucetta (Lacépède, 1803)	

TABLE I
(Continued)

Argulus species	Country/city/town	Water system as published	Water system confirmed[a]	E	Host species	Host species confirmed[b]	Reference
	Racine County, Wisconsin, U.S.A.	Tichigan Lake	Tichigan Lake	F	Cyprinus carpio	Cyprinus carpio Linnaeus, 1758	Amin (1981)
				F	Notemigonus crysoleucas	Notemigonus crysoleucas (Mitchill, 1814)	
	Ontario, Canada	Lake Huron	Lake Huron	F	Catostomus commersoni Lacépède	Catostomus commersonii (Lacépède, 1803)	Dechtiar et al. (1988)
				F	Phoxinus neogaeus Cope-finescale dace	Phoxinus neogaeus Cope, 1867	
				F	Moxostoma macrolepidotum (LeSueur) Shorthead redhorse	Moxostoma macrolepidotum (Lesueur, 1817)	
				F	Catostomus catastomus (Forster) Longnose sucker	Catostomus catostomus (Forster, 1773)	
	Ontario, Canada	Lake Erie	Lake Erie	F	Carpiodes cyprinus (LeSueur) quillback	Carpiodes cyprinus (Lesueur, 1817)	Dechtiar & Nepzy (1988)
				F	Catostomus commersoni Lacépède White sucker	Catostomus commersonii (Lacépède, 1803)	
	Ontario, Canada	Lake Ontario	Lake Ontario	F	Catostomus commersoni Lacépède White sucker	Catostomus commersonii (Lacépède, 1803)	Dechtiar & Christie (1988)

TABLE I
(Continued)

Argulus species	Country/city/town	Water system as published	Water system confirmed[a]	E	Host species	Host species confirmed[b]	Reference
	Williamson County, Illinois, U.S.A.	Lake Creek	Lake Creek	F	Lepomis gibbosus (Linnaeus) pumpkinseed	Lepomis gibbosus (Linnaeus, 1758)	Poly (1998a)
				F	Ictiobus cyprinellus	Ictiobus cyprinellus (Valenciennes, 1844)	
	Illinois, U.S.A.			F	Cyprinid host	Cyprinid	
		Fairbury	Indian Creek	F	Cyprinus carpio	Cyprinus carpio Linnaeus, 1758	
	Illinois, U.S.A.	Cache River	Cache River	B	Ictiobus bubalus (Rafinesque, 1818)	Ictiobus bubalus (Rafinesque, 1818)	
Argulus chesapeakensis Cressey, 1971[e]	Solomons, Maryland	Chesapeake biological laboratory	Patuxent River	B	Opsanus tau (Linn.) toadfish	Opsanus tau (Linnaeus, 1766)	Cressey (1971)
	Maryland to North Carolina	Central east coast of the U.S.A.	Atlantic Ocean	M	Opsanus	Opsanus sp.	Cressey (1972)
				M	Anguilla	Anguilla sp.	
				M	Rachycentron	Rachycentron canadum (Linnaeus, 1766)	
				M	Paralichthys	Paralichthys sp.	
	Chesapeake Bay south to Sapelo Island, Ga.	Southeast coast of the U.S.A.	Atlantic Ocean	M	Opsanus tau	Opsanus tau (Linnaeus, 1766)	Cressey (1978)
				M	Archosargus probatocephalus	Archosargus probatocephalus (Walbaum, 1792)	

TABLE I
(Continued)

Argulus species	Country/city/town	Water system as published	Water system confirmed[a]	E	Host species	Host species confirmed[b]	Reference
				M	Arius felis	Ariopsis felis (Linnaeus, 1766)	
				M	Mugil cephalus	Mugil cephalus Linnaeus, 1758	
				M	Gobiosoma bosci	Gobiosoma bosc (Lacépède, 1800)	
				M	Paralichthys dentatus	Paralichthys dentatus (Linnaeus, 1766)	
				M	Dasyatis americana	Dasyatis americana Hildebrand & Schroeder (1928)	
				M	Pteroplatea maclura	Gymnura micrura (Bloch & Schneider, 1801)	
				M	Rachycentron canadum	Rachycentron canadum (Linnaeus, 1766)	
Argulus chromidis Krøyer, 1863[f]	Quintana Roo, Yucatan Peninsula, Mexico	Cenote Chico	Cenote Chico	F	Rhamdia sp.	Rhamdia sp.	Wilson (1936a)
Argulus dactylopteri Thorell, 1864	West Indies		North Atlantic Ocean	M	Dactylopterus volitans (Linn.)	Dactylopterus volitans (Linnaeus, 1758)	Thorell (1864)
Argulus diversus Wilson, 1944	Not given	Not given	Not given	U	Not given	Not given	Wilson (1944)
	Bladen County, North Carolina	White Lake	White Lake	F	Ictalurus natalis (LeSueur) yellow bullhead	Ameiurus natalis (Lesueur, 1819)	Yeatman (1965)

TABLE I
(Continued)

Argulus species	Country/city/town	Water system as published	Water system confirmed[a]	E	Host species	Host species confirmed[b]	Reference
	Indiana	Not given	Not given	F	Ameriurus	Ameiurus sp.	Cressey (1972)
	Lumberton, North Carolina	Sand Hole	Lumber River	F	Ictalurus melas (Rafinesque) black bullhead	Ameiurus melas (Rafinesque, 1820)	Knuckles (1972)
Argulus flavescens Wilson, 1916	Fairport, Muscatine County, Iowa, U.S.A.	Sunfish Lake	Sunfish Lake, Mississippi River	F	Amia calva	Amia calva Linnaeus, 1766	Wilson (1916)
	Fairport, Muscatine County, Iowa, U.S.A.	Fairport	Mississippi River	F	Leptops olivaris	Pylodictis olivaris (Rafinesque, 1818)	
	Sarasota, Florida, U.S.A.	Myakka River	Myakka River	B	Aplites salmoides	Micropterus salmoides (Lacépède, 1802)	Mueller (1936)
	Clewiston, Florida, U.S.A.	Lake Okeechobee	Lake Okeechobee	F	Amia calva	Amia calva Linnaeus, 1766	
	Southern Florida, U.S.A.	Lake Okeechobee and surrounding water systems	Lake Okeechobee	F	Amia calva	Amia calva Linnaeus, 1766	Bangham (1940)
				F	Erimyson sucetta sucetta	Erimyzon sucetta (Lacépède, 1803)	
				F	Ictalurus lacustris punctatus	Ictalurus punctatus (Rafinesque, 1818)	
				F	Ameiurus nebulosus marmoratus	Ameiurus nebulosus (Lesueur, 1819)	
				F	Ameiurus natilis	Ameiurus natalis (Lesueur, 1819)	

TABLE I
(Continued)

Argulus species	Country/city/town	Water system as published	Water system confirmed[a]	E	Host species	Host species confirmed[b]	Reference
				F	*Floridichthys carpio carpio*	*Floridichthys carpio* (Günther, 1866)	
	Bryan County, Oklahoma, U.S.A. southeastern U.S.	Lake Texoma	Lake Texoma	F	*Cyprinus carpio* L.	*Cyprinus carpio* Linnaeus, 1758	Roberts (1957)
		Mississippi River system to the Coastal Gulf of Mexico	Mississippi River system to the Coastal Gulf of Mexico (Atlantic Ocean)	B	*Amia*	*Amia* sp.	Cressey (1972)
				B	*Micropterus*	*Micropterus* sp.	
				B	*Micropogon*	*Micropogon* sp.	
				B	*Paralichthys*	*Paralichthys* sp.	
				B	*Mugil*	*Mugil* sp.	
				B	*Dasyatis*	*Dasyatis* sp.	
	Guernsey County, Ohio, U.S.A.	Wills Creek	Wills Creek	F	*Ictalurus punctatus*	*Ictalurus punctatus* (Rafinesque, 1818)	Poly (1997)
	Williamson County, Illinois, U.S.A.	Lake Creek	Lake Creek	F	*Micropterus salmoides*	*Micropterus salmoides* (Lacépède, 1802)	Poly (1998a)
	Illinois, U.S.A.	Mississippi River (Pool 26)	Mississippi River (Pool 26)	F	*Notropis wickliffi*	*Notropis wickliffi* Trautman, 1931	
Argulus floridensis Meehan, 1940[g]	Key West, Monroe County, Florida		Gulf of Mexico, Atlantic Ocean	M	Host unknown	Host unknown	Meehean (1940)
	Gulf Coast of the U.S.A.		Atlantic Ocean	M	*Mugil*	*Mugil* sp.	Cressey (1972)

TABLE I
(Continued)

Argulus species	Country/city/town	Water system as published	Water system confirmed[a]	E	Host species	Host species confirmed[b]	Reference
Argulus funduli Krøyer, 1863	New Orleans, Louisiana, U.S.A.	Not given	Not given	U	*Fundulus lumbatus*	*Fundulus pulvereus* (Evermann, 1892)	Krøyer (1863)
	Massachusetts, U.S.A.	Waquoit	Waquoit Bay, Atlantic Ocean	M	Not given	Not given	Wilson (1902)
		Long Island Sound	Long Island Sound, Atlantic Ocean	M	Not given	Not given	
	Massachusetts, U.S.A.	Woods Hole	Woods Hole, North Atlantic Ocean	M	Not given	Not given	
	Beaufort, Carteret County, North Carolina	Unknown	Atlantic Ocean	M	*Fundulus heteroclitus*	*Fundulus heteroclitus* (Linnaeus, 1766)	Wilson (1907)
	Massachusetts, U.S.A.	Woods Hole	Woods Hole, North Atlantic Ocean	M	*Fundulus heteroclitus*	*Fundulus heteroclitus* (Linnaeus, 1766)	Wilson (1932)
				M	*Fundulus majalis*	*Fundulus majalis* (Walbaum, 1792)	
	Gulf of Mexico, Florida	Lemon Bay	Lemon Bay, Atlantic Ocean	M	*Lagodon rhomboides* pinfish	*Lagodon rhomboides* (Linnaeus, 1766)	Bere (1936)
	Maine to Mississippi	East coast of the U.S.A.	North Atlantic Ocean	M	*Fundulus*	*Fundulus* sp.	Cressey (1972)
				M	*Lagodon*	*Lagodon* sp.	
				M	*Chaetodon*	*Chaetodon* sp.	

TABLE I
(Continued)

Argulus species	Country/city/town	Water system as published	Water system confirmed[a]	E	Host species	Host species confirmed[b]	Reference
	New Brunswick, Canada, south to the mouth of the Neches River, Texas	East coast of North America	North Atlantic Ocean	M	*Menidia notata*	*Menidia menidia* (Linnaeus, 1766)	Cressey (1978)
				M	*Lagodon rhomboides*	*Lagodon rhomboides* (Linnaeus, 1766)	
				M	*Pseudopleuronectes americanus*	*Pseudopleuronectes americanus* (Walbaum, 1792)	
				M	*Cyprinodon variegatus*	*Cyprinodon variegatus variegatus* Lacépède, 1803	
				M	*Fundulus grandis*	*Fundulus grandis* Baird & Girard, 1853	
				M	*Fundulis heteroclitus*	*Fundulus heteroclitus* (Linnaeus, 1766)	
				M	*Fundulus majalis*	*Fundulus majalis* (Walbaum, 1792)	
				M	*Fundulus ocellaris*	*Fundulus confluentus* Goode & Bean, 1879	

TABLE I
(Continued)

Argulus species	Country/city/town	Water system as published	Water system confirmed[a]	E	Host species	Host species confirmed[b]	Reference
	Isle Vert, Quebec	Spartina patens zone of salt marsh (48°01'N 69°21'W)	salt marsh (48°01'N 69°21'W)	B	Gasterosteus aculeatus	Gasterosteus aculeatus Linnaeus, 1758	Poulin (1999)
Argulus fuscus Bere, 1936	Gulf of Mexico, Florida	Lemon Bay	Lemon Bay, Atlantic Ocean	B	Gasterosteus wheatlandi	Gasterosteus wheatlandi Putnam, 1867	Bere (1936)
				M	Orthopristis chrysopterus hogfish	Orthopristis chrysoptera (Linnaeus, 1766)	
				M	Bairdiella chrysura silver perch	Bairdiella chrysoura (Lacépède, 1802)	
	Gulf Coast of the U.S.A.		Atlantic Ocean	M	Orthopristis	Orthopristis sp.	Cressey (1972)
				M	Trachinotus	Trachinotus sp.	
Argulus ingens Wilson, 1912[h]	Coahoma County, Mississippi, U.S.A.	Moon Lake	Moon Lake	F	Lepisosteus tristachus (Bloch & Schneider) alligator gar	Atractosteus tristoechus (Bloch & Schneider, 1801)	Wilson (1912b)
Argulus intectus Wilson, 1944	Massachusetts, U.S.A.	Woods Hole	Woods Hole, North Atlantic Ocean	M	Stenotomus chrysops	Stenotomus chrysops (Linnaeus, 1766)	Wilson (1944)
Argulus japonicus Thiele, 1900[ijk]	Entire U.S.		Entire U.S.A.	F	Carassius goldfish	Carassius auratus (Linnaeus, 1758)	Wilson (1944); Cressey (1972)

TABLE I
(Continued)

Argulus species	Country/city/town	Water system as published	Water system confirmed[a]	E	Host species	Host species confirmed[b]	Reference
	Entire United States		Entire U.S.A.	F	Carassius auratus	Carassius auratus (Linnaeus, 1758)	Cressey (1978)
	Racine County, Wisconsin, U.S.A.	Tichigan Lake	Tichigan Lake	F	Ictalurus punctatus (Raf.)	Ictalurus punctatus (Rafinesque, 1818)	Amin (1981)
	Brown County, Wisconsin	Fox River	Fox River	F	carp	Cyprinus carpio Linnaeus, 1758	LaMarre & Cochran (1992)
Argulus latus Smith, 1873[1]	Massachusetts, U.S.A.	Vineyard Sound	Atlantic Ocean	M	Free swimming	Free swimming	Smith (1873)
	Massachusetts, U.S.A.	Vineyard Sound	Atlantic Ocean	M	Free swimming	Free swimming	Wilson (1902)
	Massachusetts, U.S.A.	Vineyard Sound	Atlantic Ocean	M	Free swimming	Free swimming	Wilson (1932)
	Edgartown, Marthas Vineyard, Massachusetts, U.S.A.	Chappaquiddick Island	Atlantic Ocean	M	Free swimming	Free swimming	Wilson (1944)
Argulus laticauda Smith, 1873	Massachusetts, U.S.A.	Vineyard Sound	Atlantic Ocean	M	Free swimming	Free swimming	Smith (1873)
	Not given	Not given	Not given	M	Anguilla chrysypa Rafinesque	Anguilla japonica Temminck & Schlegel, 1846	Wilson (1902)
	Edgartown, Marthas Vineyard, Massachusetts, U.S.A.	Katama Bay	Katama Bay	M	Anguilla chrysypa Rafinesque	Anguilla japonica Temminck & Schlegel, 1846	

TABLE I
(Continued)

Argulus species	Country/city/town	Water system as published	Water system confirmed[a]	E	Host species	Host species confirmed[b]	Reference
	Not given	Not given	Not given	M	Pleuronectes americanus Walbaum flatfish	Pseudopleuronectes americanus (Walbaum, 1792)	
	Edgartown, Marthas Vineyard, Massachusetts, U.S.A.	Katama Bay	Katama Bay	M	Pleuronectes americanus Walbaum flatfish	Pseudopleuronectes americanus (Walbaum, 1792)	
	Massachusetts, U.S.A.	Woods Hole	Woods Hole, North Atlantic Ocean	M	Pleuronectes americanus Walbaum flatfish	Pseudopleuronectes americanus (Walbaum, 1792)	
				M	Blenny	Blenny	
				M	Skate	skate	
				M	Sculpin	Hemitripterus americanus (Gmelin, 1789)	
				M	Bonnet skate	Aetobatus narinari (Euphrasen, 1790)	
				M	Microgadus tomcod	Microgadus tomcod (Walbaum, 1792)	
				M	Paralichthys dentatus	Paralichthys dentatus (Linnaeus, 1766)	
	Massachusetts, U.S.A.	Woods Hole	Woods Hole, North Atlantic Ocean	M	Opsanus tau	Opsanus tau (Linnaeus, 1766)	Wilson (1924)
	Massachusetts, U.S.A.	Woods Hole	Woods Hole, North Atlantic Ocean	M	Anguilla rostrata	Anguilla rostrata (Lesueur, 1817)	Wilson (1932)

TABLE I
(Continued)

Argulus species	Country/city/town	Water system as published	Water system confirmed[a]	E	Host species	Host species confirmed[b]	Reference
				M	Pseudopleuronectes americanus	Pseudopleuronectes americanus (Walbaum, 1792)	
				M	Paralichthys dentatus	Paralichthys dentatus (Linnaeus, 1766)	
				M	Microgadus tomcod	Microgadus tomcod (Walbaum, 1792)	
				M	Myoxocephalus scorpius	Myoxocephalus scorpius (Linnaeus, 1758)	
	Dry Tortugas, Monroe County, Florida		Dry Tortugas Islands, Atlantic Ocean	M	Promicrops itaira	Epinephelus itajara (Lichtenstein, 1822)	Wilson (1935a)
	Gulf of Mexico, Florida	Lemon Bay	Lemon Bay, Atlantic Ocean	M	Amphotistius say stingaree	Dasyatis say (Lesueur, 1817)	Bere (1936)
				M	Pteroplatea maclura butterfly ray	Gymnura micrura (Bloch & Schneider, 1801)	
				M	Opsanus tau toadfish	Opsanus tau (Linnaeus, 1766)	
	Oxford, Talbot County, Maryland	Tred Avon River	Tred Avon River	F	10 year old boy's eye	10 year old boy's eye	Hargis (1958)
				F	Opsanus tau toadfish	Opsanus tau (Linnaeus, 1766)	
	Maryland, U.S.A.	Solomons	Partuxent River	B	Opsanus tau toadfish	Opsanus tau (Linnaeus, 1766)	Dutcher & Schwartz (1962)

TABLE I
(Continued)

Argulus species	Country/city/town	Water system as published	Water system confirmed[a]	E	Host species	Host species confirmed[b]	Reference
	Northeast coast of the U.S.A.		North Atlantic Ocean	M	*Opsanus*	*Opsanus* sp.	Cressey (1972)
				M	*Prionotus*	*Prionotus* sp.	
				M	*Pseudopleuronectes*	*Pseudopleuronectes*	
				M	*Anguilla*	*Anguilla* sp.	
				M	*Conger*	*Conger* sp.	
				M	"Sculpin"	*Myoxocephalus scorpius* (Linnaeus, 1758)	
	New England south to Long Island Sound	northeast coast of the United States	North Atlantic Ocean	M	*Prionotus* sp.	*Prionotus* sp.	Cressey (1978)
				M	*Anguilla rostrata*	*Anguilla rostrata* (Lesueur, 1817)	
				M	*Pseudopleuronectes americanus*	*Pseudopleuronectes americanus* (Walbaum, 1792)	
Argulus lepidostei Kellicott, 1877	Buffalo, New York	Niagara River	Niagara River	F	*Lepidosteus osseus*	*Lepisosteus osseus* (Linnaeus, 1758)	Kellicott (1877)
	Fairport, Muscatine County, Iowa, U.S.A.	Fairport	Mississippi River	F	*Lepisosteus platostomus*	*Lepisosteus platostomus* Rafinesque, 1820	Wilson (1916)
	Defiance, Defiance County, Ohio, U.S.A.		Auglaize River	F	*Lepidosteus osseus*	*Lepisosteus osseus* (Linnaeus, 1758)	
	Southern Florida, U.S.A.	Lake Okeechobee and surrounding water systems	Lake Okeechobee	F	*Lepisosteus platyrhinus* De Kay	*Lepisosteus platyrhincus* DeKay, 1842	Bangham (1940)

TABLE I
(Continued)

Argulus species	Country/city/town	Water system as published	Water system confirmed[a]	E	Host species	Host species confirmed[b]	Reference
	Tenessee, U.S.A.	Reelfoot Lake	Reelfoot Lake	F	*Cylindrosteus platostomus* (Rafinesque) Short-nosed Gar	*Lepisosteus platostomus* Rafinesque, 1820	Bangham & Venard (1942)
	United States	Mississippi River system and Gulf coast of Florida	Atlantic Ocean	B	*Lepisosteus*	*Lepisosteus* sp.	Cressey (1972)
	Illinois, U.S.A.	Wabash River	Wabash River	F	*Lepisosteus platostomus*	*Lepisosteus platostomus* Rafinesque, 1820	Poly (1998a)
	Williams County, Illinois, U.S.A.	Lake Creek	Lake Creek	F	*Lepisosteus platostomus*	*Lepisosteus platostomus* Rafinesque, 1820	
Argulus longicaudatus Wilson, 1944	Denton County, Texas	Lake Dallas	Lake Dallas	F	*Pomoxis annularis*	*Pomoxis annularis* Rafinesque, 1818	Wilson (1944)
Argulus lunatus Wilson, 1944	Norfolk, Virginia		Atlantic Ocean	B	*Carassius auratus*	*Carassius auratus* (Linnaeus, 1758)	Wilson (1944)
Argulus maculosus Wilson, 1902[m]	Not given	Not given	Not given	U	Not given	Not given	Wilson (1902)
	Clayton, New York	St. Lawrence River	St. Lawrence River	F	*Esox nobilior* Thompson	*Esox nobilior* Thompson	
	Culver, Indiana, Marshall County, U.S.A.	Lake Maxinkuckee	Lake Maxinkuckee	F	*Ambloplites rupestris*	*Ambloplites rupestris* (Rafinesque, 1817)	Wilson (1907, 1916)
				F	*Ameiurus natalis*	*Ameiurus natalis* (Lesueur, 1819)	

TABLE I
(Continued)

Argulus species	Country/city/town	Water system as published	Water system confirmed[a]	E	Host species	Host species confirmed[b]	Reference
				F	*Ameiurus nebulosus*	*Ameiurus nebulosus* (Lesueur, 1819)	
	Minnesota-Wisconsin, U.S.A.	Lake Pepin	Lake Pepin	F	yellow bullhead	*Ameiurus natalis* (Lesueur, 1819)	Pearse (1924)
	Southern Florida, U.S.A.	Lake Okeechobee and surrounding water systems	Lake Okeechobee	F	*Erimyson sucetta sucetta*	*Erimyzon sucetta* (Lacépède, 1803)	Bangham (1940)
	Michigan, Indiana, Missouri, Iowa, Louisianna, New York	Not given	Not given	F	*Amia*	*Amia* sp.	Cressey (1972)
				F	*Esox*	*Esox* sp.	
				F	*Umbra*	*Umbra* sp.	
	Lumberton, North Carolina	Sand Hole	Lumber River	F	*Lepomis macrochirus* Rafinesque blue gill	*Lepomis macrochirus* Rafinesque, 1819	Knuckles (1977)
	eastern half of the United States and southeastern Canada as far west as Iowa	Not given	Not given	F	*Ictalurus natilis*	*Ameiurus natalis* (Lesueur, 1819)	Cressey (1978)
				F	*Ictalurus nebulosus*	*Ameiurus nebulosus* (Lesueur, 1819)	
				F	*Amia calva*	*Amia calva* Linnaeus, 1766	
				F	*Ambloplites rupestris*	*Ambloplites rupestris* (Rafinesque, 1817)	
				F	*Erimyzon sucetta*	*Erimyzon sucetta* (Lacépède, 1803)	

TABLE I
(Continued)

Argulus species	Country/city/town	Water system as published	Water system confirmed[a]	E	Host species	Host species confirmed[b]	Reference
				F	Esox sp.	Esox sp.	
				F	Umbra limi	Umbra limi (Kirtland, 1840)	
				F	Lepisosteus osseus	Lepisosteus osseus (Linnaeus, 1758)	
	Lumberton, North Carolina	Sand Hole	Lumber River	F	Chaenobryttus coronarius (Bartram)	Lepomis gulosus (Cuvier, 1829)	Knuckles (1978)
	Lumberton, North Carolina	Sand Hole	Lumber River	F	Lepomis gibbosus (Linnaeus) pumpkinseed	Lepomis gibbosus (Linnaeus, 1758)	Knuckles (1980)
	Lumberton, North Carolina	Sand Hole	Lumber River	F	Redear sunfish	Lepomis microlophus (Günther, 1859)	Knuckles (1984)
	Lumberton, North Carolina	Jacob Swamp	Jacob Swamp	F	Lepomis microlophus (Gunther) shell cracker	Lepomis microlophus (Günther, 1859)	
Argulus meehani Cressey, 1971[n]	Everglades National Park, Florida	Royal Palm Pond	Pond at Royal Palm center	B	Lepisosteus platyrhincus Florida gar	Lepisosteus platyrhincus DeKay, 1842	Cressey (1971)
	Florida	Not given	Not given	F	Lepisosteus	Lepisosteus sp.	Cressey (1972)
Argulus megalops Smith, 1873	Massachusetts, U.S.A.	Vineyard Sound	Atlantic Ocean	M	Free swimming	Free swimming	Smith (1873)
	Massachusetts, U.S.A.	Woods Hole	Woods Hole, North Atlantic Ocean	M	Pseudopleuronectes americanus	Pseudopleuronectes americanus (Walbaum, 1792)	Wilson (1902)

TABLE I
(Continued)

Argulus species	Country/city/town	Water system as published	Water system confirmed[a]	E	Host species	Host species confirmed[b]	Reference
	Dukes County, Massachusetts, U.S.A.	Menimsha	Menemsha Pond	M	*Pseudopleuronectes americanus*	*Pseudopleuronectes americanus* (Walbaum, 1792)	
	Barnstable County, Massachusetts, U.S.A.	Cape Cod	Atlantic Ocean	M	*Pseudopleuronectes americanus*	*Pseudopleuronectes americanus* (Walbaum, 1792)	
	Massachusetts, U.S.A.	Woods Hole	Woods Hole, North Atlantic Ocean	M	*Hippoglossoides platessoides* Fabricius	*Hippoglossoides platessoides* (Fabricius, 1780)	
				M	*Paralichthys dentatus* Linnaeus	*Paralichthys dentatus* (Linnaeus, 1766)	
				M	*Lophopsetta maculata* Mitchell Spotted Flounder	*Hippoglossina oblonga* (Mitchill, 1815)	
				M	*Prionotus carolinus* Linnaeus	*Prionotus carolinus* (Linnaeus, 1771)	
				M	*Myoxocephalus octodecimspinosus* Mitchell	*Myoxocephalus octodecemspinosus* (Mitchill, 1814)	
				M	*Lophius piscatorius* Linnaeus	*Lophius piscatorius* Linnaeus, 1758	
				M	Minnow	*Phoxinus phoxinus* (Linnaeus, 1758)	
				M	Flounder	*Pseudopleuronectes americanus* (Walbaum, 1792)	

TABLE I
(Continued)

Argulus species	Country/city/town	Water system as published	Water system confirmed[a]	E	Host species	Host species confirmed[b]	Reference
				Free swimming	Free swimming	Free swimming	Wilson (1932)
	Massachusetts, U.S.A.	Woods Hole	Woods Hole, North Atlantic Ocean	M	Pseudopleuronectes americanus	Pseudopleuronectes americanus (Walbaum, 1792)	
				M	Paralichthys dentatus	Paralichthys dentatus (Linnaeus, 1766)	
				M	Hippoglossoides platessoides	Hippoglossoides platessoides (Fabricius, 1780)	
				M	Lophopsetta maculata	Hippoglossina oblonga (Mitchill, 1815)	
				M	Prionotus carolinus	Prionotus carolinus (Linnaeus, 1771)	
				M	Myoxocephalus octodecimspinosus	Myoxocephalus octodecimspinosus (Mitchill, 1814)	
				M	Lophius piscatorius	Lophius piscatorius Linnaeus, 1758	
	Puerto Penasco, Sonora, Mexico	Cholla Bay	Pacific Ocean	M	Astroscopus zephyreus Stargazer	Astroscopus zephyreus Gilbert & Starks, 1897	Causey (1960)
	Alvarado, Veracruz, Mexico	Fish Market	Gulf of Mexico, Atlantic Ocean	B	Centropomus undecimalis snook	Centropomus undecimalis (Bloch, 1792)	
	Cleveland, Ohio, U.S.A.	Aquarium	Aquarium	M	Hippocampus hudsonius	Hippocampus erectus Perry, 1810	Davis (1965)

TABLE I
(Continued)

Argulus species	Country/city/town	Water system as published	Water system confirmed[a]	E	Host species	Host species confirmed[b]	Reference
	east coast of the U.S.A. from Massachusetts to Florida		Atlantic Ocean	M	*Chilomycterus*	*Chilomycterus* sp.	Cressey (1972)
				M	*Ogcocephalus*	*Ogcocephalus* sp.	
				M	*Synodus*	*Synodus* sp.	
				M	*Prionotus*	*Prionotus* sp.	
				M	*Tautoga*	*Tautoga* sp.	
				M	*Aleutera*	*Aleuterus* sp.	
				M	*Lophius*	*Lophius* sp.	
				M	*Paralichthys*	*Paralichthys* sp.	
				M	*Microgadus*	*Microgadus* sp.	
				M	*Raia*	*Raja* sp.	
	East coast of North America from New Brunswick, Canada, to Florida		Atlantic Ocean	M	Not listed (16 species)	Not listed (16 species)	Cressey (1978)
Argulus megalops spinosus Wilson, 1944°	Cape Tormentine, New Brunswick	Gulf of St. Lawrence	Gulf of St. Lawrence, Atlantic Ocean	B	*Liopsetta putnami*	*Pleuronectes putnami* (Gill, 1864)	Wilson (1944)
				B	*Acanthocottus octodecimspinosus*	*Myoxocephalus octodecemspinosus* (Mitchill, 1814)	
Argulus melanostictus Wilson, 1935	California, United States	Monterey Bay	Pacific Ocean	M	Free swimming	Free swimming	Wilson (1935b, 1944)
	California, United States		Pacific Ocean	M	Host unknown	Host unknown	Cressey (1972)

TABLE I
(Continued)

Argulus species	Country/city/town	Water system as published	Water system confirmed[a]	E	Host species	Host species confirmed[b]	Reference
	Baja California del Norte, Mexico	Estero Beach, Bahia de Todos Santos	Bahia de Todos Santos, Pacific Ocean	M	*Leuresthes tenuis* California grunion	*Leuresthes tenuis* (Ayres, 1860)	Olson (1972)
	San Diego County	Del Mar	Del Mar, Pacific Ocean	M	*Leuresthes tenuis* California grunion	*Leuresthes tenuis* (Ayres, 1860)	Benz et al. (1995)
	Monterey, California	nearshore Pacific waters	Pacific Ocean	M	*Leuresthes tenuis* California grunion	*Leuresthes tenuis* (Ayres, 1860)	Pineda et al. (1995)
Argulus mexicanus Pineda, Páramo & Del Río, 1995	Villahermosa City, Tabasco State, Mexico	Laguna del Horizonte	Laguna Horizonte	F	*Atractosteus tropicus*	*Atractosteus tropicus* Gill, 1863	
	Centro de Estudios Agropiscicolas, Tucta Nacajuca	Aquaculture system	Aquaculture system	F	*Atractosteus tropicus*	*Atractosteus tropicus* Gill, 1863	
	Juarez Autonoma Universidad de Tabasco	Aquaculture system	Aquaculture system	F	*Atractosteus tropicus*	*Atractosteus tropicus* Gill, 1863	
	Centro de Convivencia Infantil Villahermosa, Tabasco	Zoo aquaria	Zoo aquaria	F	*Atractosteus tropicus*	*Atractosteus tropicus* Gill, 1863	
	Villahermosa City, Tabasco State, Mexico	Laguna del Chiribital	Laguna Chirivital	F	*Atractosteus tropicus*	*Atractosteus tropicus* Gill, 1863	
	Huimanguillo, Tabasco State, Mexico	Laguna del Rosario	Laguna del Rosario	F	*Atractosteus tropicus*	*Atractosteus tropicus* Gill, 1863	

TABLE I
(Continued)

Argulus species	Country/city/town	Water system as published	Water system confirmed[a]	E	Host species	Host species confirmed[b]	Reference
	Balancan, Tabasco State, Mexico	San Pedro River	San Pedro River	F	Atractosteus tropicus	Atractosteus tropicus Gill, 1863	Wilson (1916)
Argulus mississippiensis Wilson, 1916	Fairport, Muscatine County, Iowa, U.S.A.	Fairport	Mississippi River	F	Lepisosieus platostomus	Lepisosteus platostomus Rafinesque, 1820	Wilson (1916)
	Iowa, United States	Not given	Not given	F	Lepisosteus	Lepisosteus sp.	Cressey (1972)
	Pike County, Petersburg, Indiana	White River	White River	F	Lepisosteus osseus Longnose gar	Lepisosteus osseus (Linnaeus, 1758)	Benda (1974)
	Pike County, Petersburg, Indiana	White River	White River	F	Lepisosteus platostomous shortnose gar	Lepisosteus platostomus Rafinesque, 1820	
	Clinton, Illinois	Lake Carlyle	Lake Carlyle	F	Lepisosteus osseus (Linnaeus) Longnose gar	Lepisosteus osseus (Linnaeus, 1758)	Price & Buttner (1980)
	Illinois, U.S.A.	Wabash River	Wabash River	F	Lepisosteus platostomous	Lepisosteus platostomus Rafinesque, 1820	Poly (1998a)
	Illinois, U.S.A.	Lake Creek	Lake Creek	F	Lepisosteus platostomous	Lepisosteus platostomus Rafinesque, 1820	
	Illinois, U.S.A.	Grand Tower, Mississippi River	Grand Tower, Mississippi River	F	Lepisosteus platostomous	Lepisosteus platostomus Rafinesque, 1820	
Argulus niger Wilson, 1902[P]	Portland, Oregon, U.S.A.	Unknown	Unknown	U	Host unknown	Host unknown	Wilson (1902, 1909)

TABLE I
(Continued)

Argulus species	Country/city/town	Water system as published	Water system confirmed[a]	E	Host species	Host species confirmed[b]	Reference
Argulus nobilis Thiele, 1904	Dallas, Texas, U.S.A.	Not given	Not given	U	Lepidosteus sp.	Lepidosteus sp.	Thiele (1904)
	Louisiana, U.S.A.	Lake Calcasieu	Calcasieu Lake	B	Lepisosteus tristoechus alligator gars	Atractosteus tristoechus (Bloch & Schneider, 1801)	Wilson (1924)
	Mississippi and Texas, United States	Not given	Not given	B	Lepisosteus	Lepisosteus sp.	Cressey (1972)
Argulus piperatus Wilson, 1920	Nova Scotia	Shubenacadie River	Shubenacadie River	F	Free swimming	Free swimming	Wilson (1920a)
Argulus pugettensis Dana, 1853	Washington, U.S.A.	Puget Sound	Puget Sound	M	Host unknown	Host unknown	Dana (1853)
	Washington, U.S.A.	Puget Sound	Puget Sound	M	Host unknown	Host unknown	Wilson (1902)
	Seattle, Washington, U.S.A.	Unknown	Union Bay, Lake Washington	B	Oncorhynchus kisutch Coho Salmon	Oncorhynchus kisutch (Walbaum, 1792)	Wilson (1909)
	California, U.S.A.	California coast	Pacific Ocean	M	Cymatogaster aggregatus	Cymatogaster aggregata Gibbons, 1854	Wilson (1924)
	Not given	Not given	Not given	U	Cymatogaster aggregatus	Cymatogaster aggregata Gibbons, 1854	Yamaguti (1963)
				U	Hyperprosopon argenteus	Hyperprosopon argenteum Gibbons, 1854	

TABLE I
(Continued)

Argulus species	Country/city/town	Water system as published	Water system confirmed[a]	E	Host species	Host species confirmed[b]	Reference
				U	*Salmo irideus*	*Oncorhynchus mykiss* (Walbaum, 1792)	
				U	*Salmo gaerdneri*	*Oncorhynchus mykiss* (Walbaum, 1792)	
				U	*Taeniotoca lateralis*	*Embiotoca lateralis* Agassiz, 1854	
				U	*Phanerodon furcatus*	*Phanerodon furcatus* Girard, 1854	
	northern west coast of the U.S.A.	northern west coast of the U.S.A.	Pacific Ocean	M	*Salmo*	*Salmo* sp.	Cressey (1972)
				M	*Embiotoca*	*Embiotoca* sp.	
				M	*Amphistichus*	*Amphistichus* sp.	
Argulus rhamdiae Wilson, 1936	Yucatán, Mexico	Scan Yui Cenote	Scan Yui Cenote	F	*Rhamdia* sp.	*Rhamdia* sp.	Wilson (1936a)
Argulus rotundus Wilson, 1944	Mexico	Gulf of Mexico	Gulf of Mexico, Atlantic Ocean	M	Host unknown	Host unknown	Wilson (1944)
Argulus stizostethii Kellicott, 1880	Buffalo, New York	Niagara River	Niagara River	F	*Stizostethium salmoneum*	*Sander vitreus* (Mitchill, 1818)	Kellicott (1880)
				F	*Lepidosteus osseus*	*Lepisosteus osseus* (Linnaeus, 1758)	
	Not given	Mississippi River	Mississippi River	F	Walleye	*Sander vitreus* (Mitchill, 1818)	Wilson (1916)

TABLE I
(Continued)

Argulus species	Country/city/town	Water system as published	Water system confirmed[a]	E	Host species	Host species confirmed[b]	Reference
				F	Sauger, S. canadense	Sander canadensis (Griffith & Smith, 1834)	
	Culver, Indiana, Marshall County, U.S.A.	Lake Maxinkuckee	Lake Maxinkuckee	F	Walleye	Sander vitreus (Mitchill, 1818)	Wilson (1924)
	Vilas County, Wisconsin, U.S.A.	Squirrel Lake	Squirrel Lake	F	Esox nobilior Thompson	Esox nobilior Thompson	
	Ohio, U.S.A.	Buckeye Lake	Buckeye Lake	F	Lepibema chrysops (Rafinesque)	Morone chrysops (Rafinesque, 1820)	Bangham (1941)
	northeast U.S.A. from Minnesota to Maine	Not given	Not given	B	Accipenser	Accipenser sp.	Cressey (1972)
				B	Alosa	Alosa sp.	
				B	Esox	Esox sp.	
				B	Notropis	Notropis sp.	
				B	Salvelinus	Salvelinus sp.	
				B	Gasterosteus	Gasterosteus sp.	
				B	Coregonus	Coregonus sp.	
				B	Salvelinus fontinalis	Salvelinus fontinalis (Mitchill, 1814)	
	north eastern quarter of the United States from Iowa and Minnesota east to New England and New Brunswick, Canada	Not given	Not given	B	Stizostedion vitreu	Sander vitreus (Mitchill, 1818)	Cressey (1978)
				B	Stizostedion canadens	Sander canadensis (Griffith & Smith, 1834)	

TABLE I
(Continued)

Argulus species	Country/city/town	Water system as published	Water system confirmed[a]	E	Host species	Host species confirmed[b]	Reference
				B	Acipenser fulvescens	Acipenser fulvescens Rafinesque, 1817	
				B	Esox masquinongy	Esox masquinongy Mitchill, 1824	
				B	Aloso sapidissima	Alosa sapidissima (Wilson, 1811)	
				B	Gasterosteus sp.	Gasterosteus sp.	
				B	Notropis sp.	Notropis sp.	
				B	Coregonus sp.	Coregonus sp.	
				B	Dorosoma cepedianum	Dorosoma cepedianum (Lesueur, 1818)	
	Williamson County, Illinois, U.S.A.	Crab Orchard Lake	Crab Orchard Lake	F	Pomoxis annularis Rafinesque White Crappie	Pomoxis annularis Rafinesque, 1818	Buttner (1980)
				F	Free swimming	Free swimming	
	Benton, Illinois, U.S.A.	Rend Lake	Rend Lake	F	Free swimming	Free swimming	
	Alexander County, Illinois, U.S.A.	Horseshoe Lake	Horseshoe Lake	F	Pomoxis annularis Rafinesque, 1818	Pomoxis annularis Rafinesque, 1818	Poly (1998a)
				F	Notropis sp.	Notropis sp.	
	Springfield, Illinois, U.S.A.	Lake Springfield	Lake Springfield	F	Alosa chrysochloris (Rafinesque, 1820)	Alosa chrysochloris (Rafinesque, 1820)	

TABLE I
(Continued)

Argulus species	Country/city/town	Water system as published	Water system confirmed[a]	E	Host species	Host species confirmed[b]	Reference
Argulus trilineatus Wilson, 1904	Jackson County, Illinois, U.S.A.	Kinkaid Lake	Kinkaid Lake	F	Free swimming	Free swimming	Wilson (1904)
	Efingham County, Illinois, U.S.A.	Lake Sara	Lake Sara	F	Free swimming	Free swimming	Wilson (1916)
	Macon, Georgia, U.S.A.	Aquarium	Aquarium	F	Goldfish	*Carassius auratus* (Linnaeus, 1758)	
	Henderson, Kentucky, U.S.A.	Aquarium	Aquarium	F	Goldfish	*Carassius auratus* (Linnaeus, 1758)	
	Seattle, Washington, U.S.A.	private pond	private pond	F	*Carassius auratus*	*Carassius auratus* (Linnaeus, 1758)	Guberlet (1928)
	Ontario, Canada	Lake Erie	Lake Erie	F	Free swimming	Free swimming	Tidd (1931)
	San Diego, California	State Teachers College aquarium	State Teachers College aquarium	F	Goldfish	*Carassius auratus* (Linnaeus, 1758)	Wilson (1935b)
	Louisiana, U.S.A.	Natchitoches	Cane River Lake	F	*Carassius auratus*	*Carassius auratus* (Linnaeus, 1758)	Meehean (1937)
	Takoma Park, Montgomery County, Maryland, U.S.A.		Sligo Creek	F	Goldfish	*Carassius auratus* (Linnaeus, 1758)	Wilson (1944)
Argulus varians Bere, 1936[f]	Gulf of Mexico, Florida, U.S.A.	Lemon Bay	Lemon Bay, Atlantic Ocean	M	*Lagodon rhomboides* pinfish	*Lagodon rhomboides* (Linnaeus, 1766)	Bere (1936)

TABLE I
(Continued)

Argulus species	Country/city/town	Water system as published	Water system confirmed[a]	E	Host species	Host species confirmed[b]	Reference
				M	*Ogcocephalus* sp. batfish	*Ogcocephalus* sp.	
				M	*Echeneis naucrates* sucking fish	*Echeneis naucrates* Linnaeus, 1758	
				M	*Chilomycterus spinosus* spiny toadfish	*Chilomycterus spinosus spinosus* (Linnaeus, 1758)	
	Miami, Florida, U.S.A.	Biscayne Bay	Biscayne Bay, Atlantic Ocean	M	*Sphoeroides testudineus* (L.) Checkered puffer	*Sphoeroides testudineus* (Linnaeus, 1758)	Bouchet (1985)
Argulus versicolor Wilson, 1902	Warren, Massachusetts, U.S.A.	Powdermill pond	Powdermill pond	F	*Lucius reticulatus* Le Sueur common pickerel	*Esox lucius* Linnaeus, 1758	Wilson (1902)
	Brookfield, Massachusetts, U.S.A.	Wickaboag pond	Wickaboag pond	F	*Lucius reticulatus* Le Sueur common pickerel	*Esox lucius* Linnaeus, 1758	
	Worcester County, Massachusetts, U.S.A.	Podunk pond	Podunk pond	F	*Lucius reticulatus* Le Sueur common pickerel	*Esox lucius* Linnaeus, 1758	
	Brookfield, Massachusetts, U.S.A.	Lake Lashaway	Lake Lashaway	F	*Lucius reticulatus* Le Sueur common pickerel	*Esox lucius* Linnaeus, 1758	
	Holyoke, Massachusetts, U.S.A.	Ashley ponds	Ashley pond	F	*Lucius reticulatus* Le Sueur common pickerel	*Esox lucius* Linnaeus, 1758	

TABLE I
(Continued)

Argulus species	Country/city/town	Water system as published	Water system confirmed[a]	E	Host species	F	Host species confirmed[b]	Reference
	Southwick, Massachusetts, U.S.A.	Congamon ponds	Congamond Lakes		Lucius reticulatus Le Sueur common pickerel	F	Esox lucius Linnaeus, 1758	Wilson (1916)
	Culver, Indiana, Marshall County, U.S.A.	Lake Maxinkuckee	Lake Maxinkuckee		Esox lucius Linnaeus	F	Esox lucius Linnaeus, 1758	
	Kenosha County, Wisconsin, U.S.A.	Silver Lake	Silver Lake		Pomoxis nigro-maculatus (LeSueur) Black Crappie	F	Pomoxis nigromaculatus (Lesueur, 1829)	Fischthal (1947)
	Washburn County, Wisconsin, U.S.A.	Birch Lake	Birch Lake		Perca flavescens (Mitchell) Yellow Perch	F	Perca flavescens (Mitchill, 1814)	Fischthal (1950)
	Dodge County, Wisconsin, U.S.A.	Beaver Dam Lake	Beaver Dam Lake		Stizostedion v. vitreum (Mitchell) Walleye	F	Sander vitreus (Mitchill, 1818)	Fischthal (1952)
	eastern U.S. from Massachusetts To Texas	Not given	Not given		Esox	F	Esox sp.	Cressey (1972)
	eastern half of the United States; Massachusetts, Maryland, Indiana, Georgia and Texas	Not given	Not given		Amblophlites "perch"	F F	Ambloplites sp. Perca sp.	Cressey (1978)
					Esox niger	F	Esox niger Lesueur, 1818	
					Amblophlites sp. "perch"	F F	Ambloplites sp. Perca sp.	

TABLE I
(Continued)

Argulus species	Country/city/town	Water system as published	Water system confirmed[a]	E	Host species	Host species confirmed[b]	Reference
Argulus yucatanus Poly, 2005	Yucatán, Mexico	Celestun Lagoon	a lagoon near Celestún	B	Chichlasoma urophthalmus (Günther) Mayan cichlid	Cichlasoma urophthalmum (Günther, 1862)	Poly (2005)

E, environment in which it is found: B, brackish water; F, freshwater; M, marine; U, unknown.

[a]) Water system confirmed.

[b]) Host species confirmed using fishbase.org.

[c]) A. biramosus Bere, 1931 synonym of A. appendiculosus Wilson, 1927 according to Meehean (1940).

[d]) A. canadensis Wilson, 1916 synonym of A. stizostethii Kellicott, 1880 according to Meehean (1940).

[e]) A. chesapeakensis Cressey 1971 has characteristics in common with A. flavescens Wilson, 1916 and A. laticauda Smith, 1873 according to Cressey (1971).

[f]) A. chromidis Krøyer, 1863 was first found in Nicaragua.

[g]) A. floridensis Meehean, 1940 was previously labelled A. pugettensis Dana, 1853 according to Meehean (1940).

[h]) A. ingens Wilson, 1912 synonym of A. nobilis Thiele, 1904 as stated by Wilson (1924) and Meehean (1940).

[i]) Japenese specimens of A. japonicus Thiele, 1900 are the same as A. trilineatus Wilson, 1904 according to Meehean (1940).

[j]) A. japonicus Thiele, 1900 synonym of A. pelucidus Wagler, 1935 according to Hoffman (1967).

[k]) Cosmopolitan distribution.

[l]) A. latus Smith, 1873 synonym of A. funduli Krøyer, 1863 according to Meehean (1940).

[m]) A. maculosus Wilson, 1902 was described from A. americanus Wilson, 1902 specimens according to Meehean (1940).

[n]) A. meehani Cressey, 1971 has characteristics in common with A. lepidostei Kellicott, 1877 and A. nobilis Thiele (1904 according to Cressey (1971).

[o]) A. megalops var. spinosus synonym of A. megalops Smith, 1873 according to Meehean (1940).

[p]) A. niger Wilson, 1902 synonym A. pugettensis Dana, 1853 according to Meehean (1940).

[q]) A. piperatus Wilson, 1920 synonym of A. flavescens Wilson, 1916 according to Meehean (1940).

[r]) A. varians Bere, 1936 synonym of A. megalops Smith, 1873 according to Meehean (1940).

TABLE II

Freshwater *Argulus* spp. recorded from the continent of South America

Argulus species	Country/city/town	Water system as published	Water system confirmed[a]	E	Host species	Host species confirmed[b]	Reference
Argulus amazonicus Malta & Santos-Silva, 1986	Manaus, Amazonas, Brazil	Lake Janauacá, Solimões River	Janauacá Lake, Solimões River	F	*Cichla ocellaris* Schneider	*Cichla ocellaris* Bloch & Schneider, 1801	Malta & Santos Silva (1986)
				F	*Cichla temensis* Humboldt	*Cichla temensis* Humboldt, 1821	
Argulus annae Schuurmans Stekhoven, 1951	Isla Apipé Grande, San Antonio, Argentina	Río Alto Paraná	Paraná River	F	*Salminus maxillosus*	*Salminus brasiliensis* (Cuvier, 1815)	Schuurmans Stekhoven (1951)
Argulus araucanus Atria 1975	Vischuquén, Curicó Province, Chile	Lago Vichuquén 72°06'W 34°50'S	Vichuquén Lake 72°06'W 34°50'S	F	Salmonid	Salmonid	Atria (1975)
	Valdivia, Valdivia Province, Chile	Valdivia River	Valdivia River	B	*Eleginops maclovinus* Valenciennes (1930)	*Eleginops maclovinus* Valenciennes (1930)	Asencio et al. (2010)
Argulus carteri Cunnington, 1931	Makthlawaiya	Paraguayan Chaco 58°19'W 23°25'S	Paraguayan Chaco 58°19'W 23°25'S	F	*Hoplias malabaricus* (Bloch)	*Hoplias malabaricus* (Bloch, 1794)	Cunnington (1931)
Argulus chicomendesi Malta & Varella, 2000	Amazonas, Brazil	Lake Janauacá, Río Solimões 3°25'S 60°13'W	Janauacá Lake, Río Solimões 3°25'S 60°13'W	F	*Colossoma macropomum* (Cuvier, 1818)	*Colossoma macropomum* (Cuvier, 1816)	Malta & Varella (2000)
				F	*Prochilodus nigricans* (Agassiz, 1829)	*Prochilodus nigricans* Spix & Agassiz, 1829	

TABLE II
(Continued)

Argulus species	Country/city/town	Water system as published	Water system confirmed[a]	E	Host species	Host species confirmed[b]	Reference
				F	*Pseudoplatystoma tigrinum* (Valenciennes, 1840)	*Pseudoplatystoma tigrinum* (Valenciennes, 1840)	
				F	*Hipophthalmus edentates* Spix, 1829	*Hypophthalmus edentatus* Spix & Agassiz, 1829	
				F	*Pygocentrus nattereri* Kner, 1860	*Pygocentrus nattereri* Kner, 1858	
				F	*Schizodon fasciatus* Agassiz, 1829	*Schizodon fasciatus* Spix & Agassiz, 1829	
	Amazonas, Brazil	Río Negro	Black River	F	*Brycon erythropterum* (Cope, 1872)	*Brycon cephalus* (Günther, 1869)	
	Amazonas, Brazil	1. Estação de aquacultura do Inpa Amazonas	1. Estação de aquacultura do Inpa Amazonas	F	*Brycon erythropterum* (Cope, 1872)	*Brycon cephalus* (Günther, 1869)	
	Pimenteiras, RO, Brazil	Río Guaporé	Guaporé River	F	*Schizodon fasciatus* (Agassiz, 1829)	*Schizodon fasciatus* Spix & Agassiz, 1829	
	Amazonas, Brazil	2. Estação do Itacoatiara	2. Estação do Itacoatiara	F	*Colossoma macropomum*	*Colossoma macropomum* (Cuvier, 1816)	

TABLE II
(Continued)

Argulus species	Country/city/town	Water system as published	Water system confirmed[a]	E	Host species	Host species confirmed[b]	Reference
	Northern Pantanal, Cáceres, Matto Grosso, Brazil	Caiçara Bay, Paraguay River between 16°05′02.8″S 57°44′22.7″W and 16°06′41.9″S 57°45′14.6″W	Caiçara Bay, Paraguay River between 16°05′02.8″S 57°44′22.7″W and 16°06′41.9″S 57°45′14.6″W	F	Pygocentrus nattereri Kner, 1860 Serrasalmus marginatus	Pygocentrus nattereri Kner, 1858 Serrasalmus marginatus Valenciennes, 1837	Fontana et al. (2012)
Argulus chilensis Martinez, 1952	Santiago, Chile	Quebrada Córdoba, El Tabo	Córdoba ravine, El Tabo	F	Austromenidia mauleanum (Steindachner, 1896)	Odontesthes regia (Humboldt, 1821)	Martínez (1952)
Argulus chromidis Krøyer, 1863	Nicaragua	Not given	Not given	F	Chromis sp.	Chromis sp.	Krøyer (1863)
	Nicaragua	Lake Nicaragua	Lake Nicaragua	F	Chromis sp.	Chromis sp.	Wilson (1902)
Argulus cubensis Wilson, 1936	Santa Clara, Cuba	Not given	Not given	F	Cichlasoma tetracanthus	Nandopsis tetracanthus (Valenciennes, 1831)	Wilson (1936b)
Argulus elongates Heller, 1857	Brazil	Unknown	Unknown	U	Host unknown	Host unknown	Heller (1857)
	Brazil	Unknown	Unknown	U	Host unknown	Host unknown	Wilson (1902)
	Brazil	Unknown	Unknown	U	Host unknown	Host unknown	Moreira (1912, 1913)

TABLE II
(Continued)

Argulus species	Country/city/town	Water system as published	Water system confirmed[a]	E	Host species	Host species confirmed[b]	Reference
	Miranda and Abobral, Pantanal, Brazil	Miranda River Basin (Miranda, Vermelho & Abobral River, Baia da Medalha, Baia Negra, Baia Platina)	Miranda River Basin (Miranda, Vermelho & Abobral River, Baia da Medalha, Baia Negra, Baia Platina)	F	*Pygocentrus nattereri* Kner, 1860	*Pygocentrus nattereri* Kner, 1858	Carvalho et al. (2003)
				F	*Serrasalmus spilopleura* Kner, 1860	*Serrasalmus spilopleura* Kner, 1858	
				F	*Serrasalmus marginatus* Valenciennes, 1847	*Serrasalmus marginatus* Valenciennes, 1837	
	Bolivia	Río Ichilo, oxbow lakes, and Río Beni	Ichilo River, oxbow lakes, and Beni River	F	*Pseudoplatystoma fasciatum* Linnaeus, 1840	*Pseudoplatystoma fasciatum* (Linnaeus, 1766)	Mamani et al. (2004)
				F	*Pseudoplatystoma tigrinum* Valenciennes, 1840	*Pseudoplatystoma tigrinum* (Valenciennes, 1840)	
Argulus ernsti Weibezahn & Cobo, 1964	Edo. Miranda, Venezuela	Aquarium Agustin Codazzi, Colinas de Carrizales	Aquarium Agustin Codazzi, Colinas de Carrizales	F	*Carassius auratus* (Linnaeus)	*Carassius auratus* (Linnaeus, 1758)	Weibezahn & Cobo (1964)

TABLE II
(Continued)

Argulus species	Country/city/town	Water system as published	Water system confirmed[a]	E	Host species	Host species confirmed[b]	Reference
Argulus hylae Lemos de Castro & Gomes-Correa, 1985[c]	Campos, Rio de Janeiro, Brazil	Lagoa Campelo	Lagoa do Campelo	F	Hyla geographica	Hypsiboas geographicus (Spix, 1824)	Lemos de Castro & Gomes Corréa (1985)
Argulus ichesi Bouvier, 1910	Buenos Aires, Argentina	Unknown	Unknown	U	Host unknown	Host unknown	Bouvier (1910)
	Argentina	Laguna Iberá	Iberá wetlands	F	Cynodon vulpinus Spix	Rhaphiodon vulpinus Spix & Agassiz, 1829	Schuurmans Stekhoven (1951)
	Buenos Aires, Argentina	Unknown	Unknown	F	Host unknown	Host unknown	da Silva (1980)
Argulus japonicus Thiele[d]	Puerto Rico	Aquaria	Aquaria	F	Oscar	Astronotus ocellatus (Agassiz, 1831)	Bunkley-Williams & Williams (1994)
				F	Goldfish	Carassius auratus (Linnaeus, 1758)	
Argulus juparanaensis Lemos de Castro, 1950	Linhares, Espirito Santo, Brazil	Lagôa Juparanã	Lake Juparanã	B	Pachyurus squamipinnis Agassiz	Pachyurus squamipennis Agassiz, 1831	Lemos de Castro (1950)
				F	Astyanax bimaculatus (L.)	Astyanax bimaculatus (Linnaeus, 1758)	
	Linhares, Espirito Santo, Brazil	Lagôa Juparanã	Lake Juparanã	F	Megalodoras sp.	Megalodoras sp.	Malta (1982b)

TABLE II
(Continued)

Argulus species	Country/city/town	Water system as published	Water system confirmed[a]	E	Host species	Host species confirmed[b]	Reference
	Brazil	Lago Janauacá, Rio Solimões	Janauacá Lake, Solimões River	F	*Pseudoplatystoma fasciatum*	*Pseudoplatystoma fasciatum* (Linnaeus, 1766)	Malta (1984)
	Miranda and Abobral, Pantanal	Miranda River Basin (Miranda, Vermelho & Abobral River, Baia da Medalha, Baia Negra, Baia Platina)	Miranda River Basin (Miranda, Vermelho & Abobral River, Baia da Medalha, Baia Negra, Baia Platina)	F	*Pygocentrus nattereri* Kner, 1860	*Pygocentrus nattereri* Kner, 1858	Carvalho et al. (2003)
				F	*Serrasalmus spilopleura* Kner, 1860	*Serrasalmus spilopleura* Kner, 1858	
				F	*Serrasalmus marginatus* Valenciennes, 1847	*Serrasalmus marginatus* Valenciennes, 1837	
	Bolivia	Río Ichilo, oxbow lakes, and Río Beni	Ichilo River, oxbow lakes, and Beni River	F	*Pseudoplatystoma fasciatum* Linnaeus, 1840	*Pseudoplatystoma fasciatum* (Linnaeus, 1766)	Mamani et al. (2004)
				F	*Pseudoplatystoma tigrinum* Valenciennes, 1840	*Pseudoplatystoma tigrinum* (Valenciennes, 1840)	

TABLE II
(Continued)

Argulus species	Country/city/town	Water system as published	Water system confirmed[a]	E	Host species	Host species confirmed[b]	Reference
Argulus multicolor Schuurmans Stekhoven, 1937	Taperinha, Brazil	Unknown	Unknown	U	Host unknown	Host unknown	Schuurmans Stekhoven (1937)
	Caracas, Venezuela	Unknown	Unknown	U	Host unknown	Host unknown	
	Mato Grosso, Brazil	Rios Kuluene-Xingu	Kuluene River-Xingu River Basin	F	Rhaphiodon vulpinus (Spix)	Rhaphiodon vulpinus Spix & Agassiz, 1829	Lemos de Castro (1949)
	Chavantina, Mato Grosso, Brazil	Rio das Mortes	Rio das Mortes		Rhaphiodon sp.	Rhaphiodon sp.	Lemos de Castro (1951)
	Edo. Bolivar, Venezuela	Rio Parguaza	Parguaza River	F	Hydrolicus scomberoides (Cuvier)	Hydrolycus scomberoides (Cuvier, 1819)	Weibezahn & Cobo (1964)
	Brazil	Lago Janauacá, Rio Solimões	Janauacá Lake, Solimões River	F	Colossoma macropomum	Colossoma macropomum (Cuvier, 1816)	Malta (1983, 1984)
				F	Serrasalmus nattereri	Pygocentrus nattereri Kner, 1858	
				F	Cichla temensis	Cichla temensis Humboldt, 1821	
				F	Geophagus jurupari	Satanoperca jurupari (Heckel, 1840)	

TABLE II
(Continued)

Argulus species	Country/city/town	Water system as published	Water system confirmed[a]	E	Host species	Host species confirmed[b]	Reference
	Northern Pantanal, Cáceres, Matto Grosso, Brazil	Caiçara Bay, Paraguay River between 16°05′02.8″S 57°44′22.7″W and 16°06′41.9″S 57°45′14.6″W	Caiçara Bay, Paraguay River between 16°05′02.8″S 57°44′22.7″W and 16°06′41.9″S 57°45′14.6″W	F	*Pygocentrus nattereri* Kner, 1860	*Pygocentrus nattereri* Kner, 1858	Fontana et al. (2012)
				F	*Serrasalmus marginatus*	*Serrasalmus marginatus* Valenciennes, 1837	
				F	*Serrasalmus maculatus*	*Serrasalmus maculatus* Kner, 1858	
Argulus nattereri Heller, 1857	Brazil	Unknown	Unknown	F	*Hydrocyon brevidens* Cuv.	*Salminus brasiliensis* (Cuvier, 1816)	Heller (1857)
	Brazil	Unknown	Unknown	F	*Hydrocyon brevidens* Cuv.	*Salminus brasiliensis* (Cuvier, 1816)	Krøyer (1863)
	Brazil	Unknown	Unknown	F	*Salmo (Hydrocyon) brevidens* Cuvier	*Salminus brasiliensis* (Cuvier, 1816)	Wilson (1902)
	Matto-Grosso, Brazil	Not given	Not given	F	*Salminus brevidens* Cuv.	*Salminus brasiliensis* (Cuvier, 1816)	Moreira (1912, 1913)

TABLE II
(Continued)

Argulus species	Country/city/town	Water system as published	Water system confirmed[a]	E	Host species	Host species confirmed[b]	Reference
	San Pedro, Argentina	Río Paraná	Paraná River	F	Pseudoplatystoma coruscans	Pseudoplatystoma coruscans (Spix & Agassiz, 1829)	Ringuelet (1943)
	Puerto Basilio, Argentina	Rio Uruguay	Uruguay River	F	Pseudoplatystoma sp. (surubí)	Pseudoplatystoma sp.	Brian (1947)
				F	Pseudoplatystoma sp. (surubí)	Pseudoplatystoma sp.	
	Entre Rios, Argentina	Ibicuycito	Arroyo Ibicuycito	F	Salminus maxillosus	Salminus brasiliensis (Cuvier, 1815)	Ringuelet (1948)
	San Pedro, Argentina	Río Paraná	Paraná River	F	Pseudoplatystoma coruscans	Pseudoplatystoma coruscans (Spix & Agassiz, 1829)	
	Entre Rios, Argentina	Salto Grande, Río Uruguay	Salto Grande Dam, Uruguay River	F	Host unknown	Host unknown	
	Bolivia	Río Ichilo, oxbow lakes, and Río Beni	Ichilo River, oxbow lakes, and Beni River	F	Pseudoplatystoma fasciatum Linnaeus, 1840	Pseudoplatystoma fasciatum (Linnaeus, 1766)	Mamani et al. (2004)
				F	Pseudoplatystoma tigrinum Valenciennes, 1840	Pseudoplatystoma tigrinum (Valenciennes, 1840)	
Argulus paranensis Ringuelet, 1943	Entre Rios, Argentina	Río Las Conchas, Río Paraná	Paraná River	F	Salminus maxillosus (dorado)	Salminus brasiliensis (Cuvier, 1816)	Ringuelet (1943, 1948)

TABLE II
(Continued)

Argulus species	Country/city/town	Water system as published	Water system confirmed[a]	E	Host species	Host species confirmed[b]	Reference
Argulus patagonicus Ringuelet, 1943	Argentina	Lago Pellegrini, Río Negro	Pellegrini Lake	F	Percichthys trucha	Percichthys trucha (Valenciennes, 1833)	Ringuelet (1943, 1948)
Argulus paulensis Wilson, 1924[e]	Itatiba, Sao Paulo, Brazil	Not given	Not given	F	"Talrirana"	Host unknown	Wilson (1924)
Argulus pestifer Ringuelet, 1948	Entre Rios, Argentina	Ibicuycito	Arroyo Ibicuycito	F	Salminus maxillosus	Salminus brasiliensis (Cuvier, 1816)	Ringuelet (1948)
	Brazil	Lago Janauacá, Rio Solimões	Janauacá Lake, Solimões River	F	Pseudoplatystoma fasciatum	Pseudoplatystoma fasciatum (Linnaeus, 1766)	Malta (1983, 1984)
				F	Pseudoplatystoma tigrinum	Pseudoplatystoma tigrinum (Valenciennes, 1840)	
	Bolivia	Río Ichilo, oxbow lakes, and Río Beni	Ichilo River, oxbow lakes, and Beni River	F	Pseudoplatystoma fasciatum Linnaeus, 1840	Pseudoplatystoma fasciatum (Linnaeus, 1766)	Mamani et al. (2004)
				F	Pseudoplatystoma tigrinum Valenciennes, 1840	Pseudoplatystoma tigrinum (Valenciennes, 1840)	
	Brazil	floodplains of Upper Paraná River	Paraná River	F	Pseudoplatystoma corruscans (Spix & Agassiz, 1829)	Pseudoplatystoma corruscans (Spix & Agassiz, 1829)	Takemoto et al. (2009)
Argulus salmini Krøyer, 1863	Minas Gerais, Brazil	Not given	Not given	F	Salmineus sp.	Salminus sp.	Krøyer (1863)

TABLE II
(Continued)

Argulus species	Country/city/town	Water system as published	Water system confirmed[a]	E	Host species	Host species confirmed[b]	Reference
	Brazil	Unknown	Unknown		Salmo	Salmo sp.	Wilson (1902)
	Matto-Grosso, Brasil	Jaurú River	Jaurú River	F	Salminus brevidens (Cuv. & Val.) dorado	Salminus brasiliensis (Cuvier, 1816)	Moreira (1912, 1913)
	Salto Alegre, Matto-Grosso, Brasil	Jaurú River	Jaurú River	F	Salminus sp. "piraputanga"	Brycon hilari (Valenciennes, 1850)	Moreira (1913)
	Río Paraná, Argentina	Río de la Plata	Río de la Plata	F	Salminus maxillosus	Salminus brasiliensis (Cuvier, 1815)	Ringuelet (1943)
	Buenos Aires, Argentina	Río de la Plata	Río de la Plata	F	Salminus maxillosus	Salminus brasiliensis (Cuvier, 1816)	Ringuelet (1948)
	Rosario, Santa Fe, Argentina	Río Paraná	Paraná River	F	Salminus maxillosus	Salminus brasiliensis (Cuvier, 1816)	
	Col. Panambí, Misiones, Argentina	Alto Uruguay	Uruguay River	F	Salminus maxillosus	Salminus brasiliensis (Cuvier, 1816)	
	Entre Rios, Argentina	Río Gualeguaychú	Gualeguaychú River	F	Salminus maxillosus	Salminus brasiliensis (Cuvier, 1816)	
	Villarica, Paraguay	Río Tebicuary	Tebicuary River	F	Salminus maxillosus	Salminus brasiliensis (Cuvier, 1816)	

TABLE II
(Continued)

Argulus species	Country/city/town	Water system as published	Water system confirmed[a]	E	Host species	Host species confirmed[b]	Reference
Argulus salmini argentinensis Brian, 1947	Pto. Basilio, Gualeguaychu	Rio Uruguay (E. Rios)	Uruguay River	F	Salminus brevidens (Cuv. & Val.) dorado	Salminus brasiliensis (Cuvier, 1816)	Brian (1947)
	Jujuy, Argentina	Lavalleja River	Lavalleja River	F	Salminus brevidens (Cuv. & Val.) dorado	Salminus brasiliensis (Cuvier, 1816)	
Argulus silvestrii Lahille, 1926[f]	Not given	Not given	Not given	F	Pseudoplatystoma corruscans	Pseudoplatystoma corruscans (Spix & Agassiz, 1829)	Lahille (1926)
Argulus spinulosus da Silva, 1980	Rio Grande do Sul, Brazil	Rio Jacuí,	Jacuí River	F	Rhamdia sp.	Rhamdia sp.	da Silva (1980)
	Rio Grande do Sul, Brazil	Rio Jacuí, Estuario do Guaíba	Guaíba Lake, Jacuí River	F	Hoplias malabaricus Bloch, 1794	Hoplias malabaricus (Bloch, 1794)	
	Sombrio, Santa Catarina, Rio Grande do Sul, Brazil		Sombrio Lake	F	Rhamdiai sp. (jundiá)	Rhamdia quelen (Quoy & Gaimard, 1824)	
	Porto Alegre, Brazil		Guaíba Lake, Jacuí River	F	Host unknown	Host unknown	
	São Jerônimo		Jacuí River	F	Host unknown	Host unknown	
	Sertão Santana		Guaíba Lake, Jacuí River	F	Host unknown	Host unknown	
	Tapes, Brazil		Guaíba Lake, Jacuí River	F	Host unknown	Host unknown	

TABLE II
(Continued)

Argulus species	Country/city/town	Water system as published	Water system confirmed[a]	E	Host species	Host species confirmed[b]	Reference
Argulus vierai Pereira Fonseca, 1939	Montevideo, Uruguay	Unknown	Unknown	F	Cnesterodon decemmaculatus (Jenyns)	Cnesterodon decemmaculatus (Jenyns, 1842)	Pereira Fonseca (1939)
Argulus violaceus Thomsen, 1925	Montevideo, Uruguay	Unknown	Unknown	F	Rhamdia quelea	Rhamdia quelen (Quoy & Gaimard, 1824)	Thomsen (1925)
	Uruguay	Unknown	Unknown	F	Pledostomus commersonii	Hypostomus commersoni Valenciennes, 1836	Meehean (1940)
		Unknown	Unknown	F	Rhamdia quelea	Rhamdia quelen (Quoy & Gaimard, 1824)	
	Buenos Aires, Argentina	Riacho del Tigre, laguna Chascomús	Laguna Chascomus	F	Hoplias malabaricus Bloch, 1794	Hoplias malabaricus (Bloch, 1794)	Ringuelet (1943)
				F	Rhamdia sapo	Rhamdia quelen (Quoy & Gaimard, 1824)	
				F	Parapimelodus valenciennesi	Parapimelodus valenciennis (Lütken, 1874)	
				F	Loricaria anus	Loricariichthys anus (Valenciennes, 1835)	
	Argentina	Mar del Plata 37°02′S 52°35W	Mar del Plata 37°02′S 52°35W	F	Merluccinus sp.	Merluccius sp.	Brian (1947)

TABLE II
(Continued)

Argulus species	Country/city/town	Water system as published	Water system confirmed[a]	E	Host species	Host species confirmed[b]	Reference
	Buenos Aires, Argentina	Laguna de Chascomús	Laguna Chascomus	F	Host unknown	Host unknown	Ringuelet (1948)
	Srra. de la Ventana, Buenos Aires, Argentina	Arroyo del Negro	Black River	F	*Salminus irideus*	*Halichoeres brasiliensis* (Bloch, 1791)	
	Monte Hermoso, Buenos Aires	Río Sauce Grande	Río Sauce Grande	F	*Rhamdia sapo*	*Rhamdia quelen* (Quoy & Gaimard, 1824)	
	Buenos Aires, Argentina	Riacho del Tigre, laguna Chascomús	Laguna Chascomus	F	*Acestrorhamphus jennynsi*	*Oligosarcus jenynsii* (Günther, 1864)	
	Entre Rios, Argentina	Ayo. Isletas, Gualeguaychú	Gualeguaychú River	F	*Hoplias malabaricus*	*Hoplias malabaricus* (Bloch, 1794)	
	Buenos Aires, Argentina	Río Salado	Salado River	F	*Rhamdia sapo*	*Rhamdia quelen* (Quoy & Gaimard, 1824)	

E, environment in which it is found: B, brackish water; F, freshwater; M, marine; U, unknown.
[a] Water system confirmed.
[b] Host species confirmed using fishbase.org.
[c] *A. hylae* Lemos de Castro & Gomes-Correa (1985) may need to be declared "nomen nudum" because of the lack of a species description.
[d] Cosmopolitan distribution.
[e] *A. paulensis* Wilson (1924) synonym of *A. salmini* Krøyer, 1863 according to Meehean (1940).
[f] *A. silvestrii* Lahille (1926) synonym of *A. nattereri* Heller, 1857 according to Ringuelet (1943, 1948).

TABLE III

Marine and freshwater *Argulus* spp. recorded from the continent of Europe

Argulus species	Country/city/town	Water system as published	Water system confirmed[a]	E	Host species	Host species confirmed[b]	Reference
Argulus appendiculosus Wilson, 1907[c]	Wareham, Dorset, England	River Frome	River Frome	F	*Micropterus salmoides* (Lacépède) Largemouth bass-introduced	*Micropterus salmoides* (Lacépède, 1802)	Kennedy (1975)
Argulus arcassonensis Cuénot, 1912	Arcachon, France	Arcachon Bay	Arcachon Bay, Bay of Biscay	F	*Leuciscus cephalus* (L.)	*Squalius cephalus* (Linnaeus, 1758)	Cuénot (1912)
				M	*Trigla cuculus*	*Chelidonichthys cuculus* (Linnaeus, 1758)	
				M	*Symphodus melops*	*Symphodus melops* (Linnaeus, 1758)	
				M	*Symphodus cinereus*	*Symphodus cinereus* (Bonnaterre, 1788)	
				M	*Balistes capriscus*	*Balistes capriscus* Gmelin, 1789	
Argulus coregoni Thorell, 1864	Sweden	Storsjön, Jemtland	Storsjön Lake	F	*Coregono lavareto* Linn.	*Coregonus lavaretus* (Linnaeus, 1758)	Thorell (1864)
				F	*Thymall vulgari* Nilss.	*Thymallus thymallus* (Linnaeus, 1758)	
		Laxsjön, Dalsland	Laxsjön lake, Dalsland	F	*Coregano lacareto*	*Coregonus lavaretus* (Linnaeus, 1758)	

TABLE III
(Continued)

Argulus species	Country/city/town	Water system as published	Water system confirmed[a]	E	Host species	Host species confirmed[b]	Reference
		Vettern Lake	Vättern lake	F	Salmo trutta Linn.	Salmo trutta Linnaeus, 1758	Roland (1963)
	France	Lunain tributary	Lunain River	F	Trutta fario L.	Salmo trutta Linnaeus, 1758	
	Thankerton	River Clyde	River Clyde	F	Salmo trutta L.	Salmo trutta Linnaeus, 1758	Campbell (1971)
	Poland	Unknown	Unkown	F	Salmonidae	Salmonidae	Kulmatycki (1938) as cited in Grabda (1971)
	Poland	Widawka River	Widawka River	F	Leuciscus cephalus (L.)	Squalius cephalus (Linnaeus, 1758)	Penczak (1972)
	Central Norway	various lakes	various lakes	F	Salmo trutta	Salmo trutta Linnaeus, 1758	Økland (1985)
				F	Coregonus lavaretus	Coregonus lavaretus (Linnaeus, 1758)	
				F	Lota lota	Lota lota (Linnaeus, 1758)	
	Central Finland	commercial fish farm	commercial fish farm	F	Rainbow trout	Oncorhynchus mykiss (Walbaum, 1792)	Pasternak et al. (2004a, b)
	Central Finland	commercial fish farm	commercial fish farm	F	Oncorhynchus mykiss rainbow trout	Oncorhynchus mykiss (Walbaum, 1792)	Hakalahti et al. (2004c)

TABLE III
(Continued)

Argulus species	Country/city/town	Water system as published	Water system confirmed[a]	E	Host species	Host species confirmed[b]	Reference
	Central Finland	commercial fish farm	commercial fish farm	F	Host unknown	Host unknown	Fenton et al. (2006)
	Central Finland	commercial fish farm	commercial fish farm	F	Rainbow trout	Oncorhynchus mykiss (Walbaum, 1792)	Bandilla (2007)
Argulus foliaceus (Linnaeus, 1758)	Europe	stagnant freshwater	stagnant freshwater	F	Host unknown	Host unknown	Linnaeus (1758)
	Poland	Various water systems in the Masurian Lake District	Masurian Lake District	F	Gasterostei trachuri	Gasterosteus aculeatus Linnaeus, 1758	Zaddach (1884)
	Britain	Faroe Channel	Faroe-Shetland Channel	F	Host unknown	Host unknown	Scott & Scott (1913)
	Berlin, Germany	Berlin aquarium pool	Berlin aquarium pool	F	Amiatus calvus	Ameiurus catus (Linnaeus, 1758)	Herter (1926)
				F	Lepidosteus osseus	Lepisosteus osseus (Linnaeus, 1758)	
	Unknown	Unknown	Unknown	F	Triturus vulgaris vulgaris (L.) smooth newt tadpole	Lissotriton vulgaris (Linnaeus, 1758)	Bower-Shore (1940)
	Stockport, Cheshire	Reddish Canal	Stockport Branch Canal	F	Stickleback	Gasterosteus aculeatus Linnaeus, 1758	
	Pohořelice, Czech Republic	Vrkoč, Starý and Novoveský, Jihlávka River	various ponds along Jihlava River	F	Mirror Carp	Cyprinus carpio Linnaeus, 1758	Bazal et al. (1969)

TABLE III
(Continued)

Argulus species	Country/city/town	Water system as published	Water system confirmed[a]	E	Host species	Host species confirmed[b]	Reference
				F	Leather Carp	*Cyprinus carpio* Linnaeus, 1758	
	Benešov near Prague, in the Central Bohemian granitic highlands	Solopyský Horní, Hejnovka Velká, Minartický Horní, U Sedlečka and Kamenný Malý	various systems in Prague	F	Common Carp	*Cyprinus carpio* Linnaeus, 1758	
	Chesire, Liverpool	Rostherne Mere	Rostherne Mere	F	*Esox lucius*	*Esox lucius* Linnaeus, 1758	Rizvi (1969)
				F	*Rutilus rutilus*	*Rutilus rutilus* (Linnaeus, 1758)	
				F	*Perca fluviatilis*	*Perca fluviatilis* Linnaeus, 1758	
	Chesire, Liverpool	Shropshire Union Canal at Backford	Shropshire Union Canal	F	*Rutilus rutilus* (L.)	*Rutilus rutilus* (Linnaeus, 1758)	Mishra & Chubb (1969)
				F	*Abramis brama* (L.)	*Abramis brama* (Linnaeus, 1758)	
				F	*Perca fluviatilis* (L.)	*Perca fluviatilis* Linnaeus, 1758	
				F	*Esox lucius* L.	*Esox lucius* Linnaeus, 1758	
	Białystok Municipal Park, Poland	ponds	ponds	F	*Gasterosteus aculeatus* L. Three-spined sticklebacks	*Gasterosteus aculeatus* Linnaeus, 1758	Czeczuga (1971)

TABLE III
(Continued)

Argulus species	Country/city/town	Water system as published	Water system confirmed[a]	E	Host species	Host species confirmed[b]	Reference
	Perugia, Italy	Lake Tresimeno	Lake Tresimeno	F	Cyprinus carpio L.	Cyprinus carpio Linnaeus, 1758	Gattaponi (1971)
				F	Tinca tinca	Tinca tinca (Linnaeus, 1758)	
	Masuria, Poland	Masurian Lake District	Masurian Lake District	F	Salmo trutta m. Fario	Salmo trutta Linnaeus, 1758	Wegener (1909) as cited in Grabda (1971)
				F	Esox lucius	Esox lucius Linnaeus, 1758	
		Lake Luknajo and Track, Masurian Lake District	Masurian Lake District	F	Carassius carassius	Carassius carassius (Linnaeus, 1758)	Grabda (1971)
				F	Scardinius erythrophthalmus	Scardinius erythrophthalmus (Linnaeus, 1758)	
		Lake Luknajo, Masurian Lake District	Masurian Lake District	F	Blicca bjoercna	Blicca bjoerkna (Linnaeus, 1758)	Grabda (1967) as cited in Grabda (1971)
		Lake Luknajo, Masurian Lake District	Masurian Lake District	F	Rutilus rutilus	Rutilus rutilus (Linnaeus, 1758)	

TABLE III
(Continued)

Argulus species	Country/city/town	Water system as published	Water system confirmed[a]	E	Host species	Host species confirmed[b]	Reference
		Lake Ransk, Masurian Lake District	Masurian Lake District	F	*Tinca tinca*	*Tinca tinca* (Linnaeus, 1758)	Kulmatycki (1938) as cited in Grabda (1971)
		Oswin Lake, Masurian Lake District	Oświn Lake	F			
		Arklity, Masurian Lake District	Masurian Lake District	F			
		White, Masurian Lake District	Masurian Lake District	F			
		Lake Luknajno, Masurian Lake District	Masurian Lake District	F			
		Lake Mamry, Masurian Lake District	Lake Mamry, Masurian Lake District	F	*Perca fluviatilis*	*Perca fluviatilis* Linnaeus, 1758	Wegener (1909) as cited in Grabda (1971)
		Oświn Lake,	Oświn Lake	F	*Esox lucius*	*Esox lucius* Linnaeus, 1758	Hugulak (1965) as cited in Grabda (1971)

TABLE III
(Continued)

Argulus species	Country/city/town	Water system as published	Water system confirmed[a]	E	Host species	Host species confirmed[b]	Reference
		Lake Luknajo	Masurian Lake District	F			
		Radom	Mleczna River	F			
		Lake Pamer, Dobskie, Czos	Masurian Lake District	F	*Abramis brama*	*Abramis brama* (Linnaeus, 1758)	Grabda (1971)
		Lake Luknajo, Oswin, Karas, Dobskie, Wuksniki, Malszewo	Masurian Lake District	F	*Perca fluviatilis*	*Perca fluviatilis* Linnaeus, 1758	Kozikowska (1961) as cited in Grabda (1971)
	Poland	Upper Silesia	Oder River	F		*Rutilus rutilus* (Linnaeus, 1758)	
	Poland	Upper Silesia: dam reservoir in Koslov Gora	Kozlowa Góra Lake	F	*Rutilus rutilus*		
	Poland	Upper Silesia: dam reservoir in Koslov Gora	Kozlowa Góra Lake	F	*Perca fluviatilis*	*Perca fluviatilis* Linnaeus, 1758	Hugulak (1965) as cited in Grabda (1971)
	Poland	Zhiornik barrage in Kozlov Gora	Kozlowa Góra Lake	F	*Esox lucius*	*Esox lucius* Linnaeus, 1758	Hugulak (1965) as cited in Grabda (1971)

TABLE III
(Continued)

Argulus species	Country/city/town	Water system as published	Water system confirmed[a]	E	Host species	Host species confirmed[b]	Reference
	Ochaby, Poland	Upper Silesia: Ponds in Ochaby	Vistula River	F	*Lucioperca lucioperca*	*Sander lucioperca* (Linnaeus, 1758)	Grabda (1967) as cited in Grabda (1971)
	Swaderkach, Poland	Ponds	Ponds, Masurian Lake District	F	*Coregonus peled* (Gmel.)	*Coregonus peled* (Gmelin, 1789)	Grabda (1967) as cited in Grabda (1971)
	West Pomerania, Poland	Pomeranian Lake District	Pomeranian Lake District	F	*Tinca tinca*	*Tinca tinca* (Linnaeus, 1758)	Wundsch (1927) as cited in Grabda (1971)
	West Pomerania, Poland	Lake Smolnik, Pomeranian Lake District	Smolnik Lake, Pomeranian Lake District	F	*Salmo trutta m. lacustris*	*Salmo trutta* Linnaeus, 1758	Grabda et al. (1961) as cited in Grabda (1971)
		Lake Wdzydze	Wdzydze Lake	F	*Esox lucius*	*Esox lucius* Linnaeus, 1758	
		Trzebiocha stream		F	*Gasterosteus aculeatus*	*Gasterosteus aculeatus* Linnaeus, 1758	Grabda (1961) as cited in Grabda (1971)

TABLE III
(Continued)

Argulus species	Country/city/town	Water system as published	Water system confirmed[a]	E	Host species	Host species confirmed[b]	Reference
	West Pomerania, Poland	Wisla River under Rafa, Nogat, Lakes marsh and Parnowo	Pomeranian Lake District	F	*Abramis brama*	*Abramis brama* (Linnaeus, 1758)	Grabda (1971)
	Wicko, Baltic Coast, Poland	Not given	Not given	F	*Perca fluviatilis*	*Perca fluviatilis* Linnaeus, 1758	Kozikowska (1957) as cited in Grabda (1971)
	Baltic Coast, Poland	Lake Jamno	Lake Jamno	F	*Esox lucius*	*Esox lucius* Linnaeus, 1758	Lehmann (1924) as cited in Grabda (1971)
				F	*Abramis brama*	*Abramis brama* (Linnaeus, 1758)	Grabda & Grabda (1957) as cited in Grabda (1971)
		Lake Druzno	Drużno Lake	F	*Abramis brama*	*Abramis brama* (Linnaeus, 1758)	Kozicka (1959) as cited in Grabda (1971)

TABLE III
(Continued)

Argulus species	Country/city/town	Water system as published	Water system confirmed[a]	E	Host species		Host species confirmed[b]	Reference
					Perca fluviatilis		*Perca fluviatilis* Linnaeus, 1758	Seligo (1900) as cited in Grabda (1971)
		Barlewickie Jezioro	Barlewickie Lake	F	*Leus caspius delineates* (Heck.)		*Leucaspius delineatus* (Heckel, 1843)	
		Zajezierskie Jezioro	Zajezierskie Lake					
		The water reservoir Zegrzynski	Zegrze Reservoi	F	*Alburnus alburnus*		*Alburnus alburnus* (Linnaeus, 1758)	Borowik (1967, 1968) as cited in Grabda (1971)
		Vistula Lagoon	Vistula Lagoon	B	*Perca fluviatilis*		*Perca fluviatilis* Linnaeus, 1758	Wegener (1909) and Wagler (1935) as cited in Grabda (1971)
	Warsaw, Poland	Wisla River	Wisla River	F	*Abramis brama*		*Abramis brama* (Linnaeus, 1758)	Dabrowska (1970) and Kozicka (1949, 1951) as cited in Grabda (1971)

TABLE III
(Continued)

Argulus species	Country/city/town	Water system as published	Water system confirmed[a]	E	Host species	Host species confirmed[b]	Reference
	West Poland	Lake Powidzkie	Powidzkie Lake	F	Blicca bjoerena	Blicca bjoerkna (Linnaeus, 1758)	Kulmatycki (1938) as cited in Grabda (1971)
				F	Rutilus rutilus	Rutilus rutilus (Linnaeus, 1758)	
				F	Perca fluviatilis	Perca fluviatilis Linnaeus, 1758	
	Poland	Białowieza Forest	Orłówka River	F	Perca fluviatilis	Perca fluviatilis Linnaeus, 1758	Wagler (1935) as cited in Grabda (1971)
	Hillerød, Denmark	Frederiksborg Castle Lake	Frederiksborg Castle Lake	F	Host unknown	Host unknown	Wingstrand (1972)
	Léman, France	Lake Geneva	Lake Geneva	F	Perca fluviatilis	Perca fluviatilis Linnaeus, 1758	Laurent (1975)
				F	Rutilus rutilus	Rutilus rutilus (Linnaeus, 1758)	
				F	Esox lucius	Esox lucius Linnaeus, 1758	
	Nakrivnje, Leskovac region, Serbia	private fishpond	private fishpond	F	Carp	Cyprinus carpio Linnaeus, 1758	Canić et al. (1977)
	Czechoslovakia	Břehyně Brook (North Bohemia, River Elbe Basin)	Břehyně Brook (North Bohemia, River Elbe Basin)	F	Host unknown	Host unknown	Moravec (1978)

TABLE III
(Continued)

Argulus species	Country/city/town	Water system as published	Water system confirmed[a]	E	Host species	Host species confirmed[b]	Reference
	Liblar, Cologne	Pond	pond	F	Xiphophorus helleri Heckel	Xiphophorus hellerii Heckel, 1848	Schlüter (1979)
	Scania, Sweden	lakes in Scania	various lakes in Scania	F	Not given	Not given	Hallberg (1982)
	Lithuania	Lake Zhialva	Lake Zhialva	F	Pelyad	Coregonus peled (Gmelin, 1789)	Zhiliukas & Rauckis (1982)
	Norway	various lakes	various lakes	F	Perca fluviatilis	Perca fluviatilis Linnaeus, 1758	Økland (1985)
	Denmark	Lake Esrum	Lake Esrum	F	Anguilla anguilla	Anguilla anguilla (Linnaeus, 1758)	Køie (1988)
	Azores	Lagoa das Sete Cidades, S. Miguel Island	Lagoa das Sete Cidades, S. Miguel Island	F	Pike	Esox lucius Linnaeus, 1758	Menezes et al. (1990)
				F	Perch	Perca fluviatilis Linnaeus, 1758	
				F	Carp	Cyprinus carpio Linnaeus, 1758	
				F	Rainbow trout	Oncorhynchus mykiss (Walbaum, 1792)	
	Wroclaw, Poland	Szczodre fish ponds	fish ponds in Szczodre	F	Not recorded	Not recorded	Kubrakiewicz & Klimowicz (1994)
	Jutland, Denmark	consumption fish, fish farm	consumption fish, fish farm	F	Oncorhynchus mykiss	Oncorhynchus mykiss (Walbaum, 1792)	Buchmann et al. (1995)

TABLE III
(Continued)

Argulus species	Country/city/town	Water system as published	Water system confirmed[a]	E	Host species	Host species confirmed[b]	Reference
	Jutland, Denmark	consumption fish, fish farm	consumption fish, fish farm	F	*Oncorhynchus mykiss*	*Oncorhynchus mykiss* (Walbaum, 1792)	Buchmann & Bresciani (1997)
	Central Scotland	stillwater fishery	stillwater fishery	F	*Oncorhynchus mykiss* (Walbaum)	*Oncorhynchus mykiss* (Walbaum, 1792)	Northcott et al. (1997)
				F	*Salmo trutta* L.	*Salmo trutta* Linnaeus, 1758	
				F	Stickleback	*Gasterosteus aculeatus* Linnaeus, 1758	
				F	*Rutilus rutilus* (L.)	*Rutilus rutilus* (Linnaeus, 1758)	
	Central Finland	Rutalahti Bay, Lake Päijänne	Rutalahti Bay, Lake Päijänne	F	*Perca fluviatilis* perch	*Perca fluviatilis* Linnaeus, 1758	Mikheev et al. (1998, 2000)
	Hungary	Lake Balaton	Lake Balaton	F	*Aspius aspius*	*Leuciscus aspius* (Linnaeus, 1758)	Molnár & Székely (1998)
		Kis-Balaton Reservoir	Kis-Balaton Reservoir	F	*Cyprinus carpio*	*Cyprinus carpio* Linnaeus, 1758	
				F	*Aspius aspius*	*Leuciscus aspius* (Linnaeus, 1758)	
				F	*Abramis brama*	*Abramis brama* (Linnaeus, 1758)	
				F	*Ctenopharyngodon idella*	*Ctenopharyngodon idella* (Valenciennes, 1844)	

TABLE III
(Continued)

Argulus species	Country/city/town	Water system as published	Water system confirmed[a]	E	Host species	Host species confirmed[b]	Reference
	Finland	Lake Päijänne to Gulf of Finland	Lake Päijänne	F	Tinca tinca	Tinca tinca (Linnaeus, 1758)	Pasternak et al. (2000)
				F	Stizostedion lucioperca	Sander lucioperca (Linnaeus, 1758)	
	Bangor, County Down, Northern Ireland	Lower Ballysallagh reservoir 54°37'N 5°45'W	Lower Ballysallagh reservoir 54°37'N 5°45'W	F	Perca fluviatilis L. perch	Perca fluviatilis Linnaeus, 1758	Gault et al. (2002)
				F	Rutilus rutilus (L.) roach	Rutilus rutilus (Linnaeus, 1758)	
				F	Oncorhynchus myskiss rainbow trout	Oncorhynchus mykiss (Walbaum, 1792)	
	Bangor, County Down, Northern Ireland	fishery 54°37'N 5°45'W	fishery 54°37'N 5°45'W	F	Oncorhynchus mykiss rainbow trout	Oncorhynchus mykiss (Walbaum, 1792)	Harrison et al. (2006)
				F	Gasterosteus aculeatus L. Three-spined sticklebacks	Gasterosteus aculeatus Linnaeus, 1758	
	Portaferry, County Down, Northern Ireland	fishery 54°24'N 05°33'W	fishery 54°24'N 05°33'W	F	Oncorhynchus mykiss rainbow trout	Oncorhynchus mykiss (Walbaum, 1792)	
				F	Esox luctus pike	Esox lucius Linnaeus, 1758	
				F	Perca fluviatilis perch	Perca fluviatilis Linnaeus, 1758	

TABLE III
(Continued)

Argulus species	Country/city/town	Water system as published	Water system confirmed[a]	E	Host species	Host species confirmed[b]	Reference
	Bangor, County Down, Northern Ireland	Lower Ballysallagh reservoir 54°37'N 5°45'W	Lower Ballysallagh reservoir 54°37'N 5°45'W	F	Scardinius erythrophthalmus rudd	Scardinius erythrophthalmus (Linnaeus, 1758)	Harrison et al. (2007)
				F	Oncorhynchus myskiss rainbow trout	Oncorhynchus mykiss (Walbaum, 1792)	
				F	Gasterosteus aculeatus L. Three-spined sticklebacks	Gasterosteus aculeatus Linnaeus, 1758	
	Charlottenlund, Denmark	Danish National Aquarium	Danish National Aquarium	F	Free swimming (Plankton)	Free swimming (Plankton)	Møller et al. (2007, 2008, 2009)
	Copenhagen, Denmark	Utterslev mose	Utterslev mose	F	Rutilus rutilus (L.)	Rutilus rutilus (Linnaeus, 1758)	Møller (2009)
				F	Abramis brama (L.)	Abramis brama (Linnaeus, 1758)	
				F	Scardinius erythropthalmus (L.)	Scardinius erythrophthalmus (Linnaeus, 1758)	
	South Wales	stillwater trout fishery	stillwater trout fishery	F	Oncorhynchus mykiss rainbow trout	Oncorhynchus mykiss (Walbaum, 1792)	Taylor et al. (2009a, b)

TABLE III
(Continued)

Argulus species	Country/city/town	Water system as published	Water system confirmed[a]	E	Host species	Host species confirmed[b]	Reference
	Gloucestershire			F	Oncorhynchus mykiss rainbow trout	Oncorhynchus mykiss (Walbaum, 1792)	
	Staffordshire			F	Oncorhynchus mykiss rainbow trout	Oncorhynchus mykiss (Walbaum, 1792)	
	Wiltshire			F	Oncorhynchus mykiss rainbow trout	Oncorhynchus mykiss (Walbaum, 1792)	
Argulus giordanii Brian, 1959[d]	Tuscany, Central Italy	Massaciuccoli water district	Massaciuccoli water district	F	Carassius auratus (Linnaeus, 1758)	Carassius auratus (Linnaeus, 1758)	Macchioni et al. (2015)
Argulus japonicus Thiele, 1900[e]	Pohořelice, Czech Republic	Vrkoč, Starý and Novoveský, Jihlávka River	various ponds along Jihlava River	F	Mirror Carp	Cyprinus carpio Linnaeus, 1758	Bazal et al. (1969)
				F	Leather Carp	Cyprinus carpio Linnaeus, 1758	
	Benešov near Prague, in the Central Bohemian granitic highlands	Solopyský Horní, Hejnovka Velká, Minartický Horní, U Sedlečka and Kamenný Malý	various systems in Prague	F	Common Carp	Cyprinus carpio Linnaeus, 1758	

TABLE III
(Continued)

Argulus species	Country/city/town	Water system as published	Water system confirmed[a]	E	Host species	Host species confirmed[b]	Reference
	Kujawska, Poland	Skepe and Kijaszkowie	Unknown	E	*Cyprinus carpio*	*Cyprinus carpio* Linnaeus, 1758	Grabda (1955, 1967) as cited in Grabda (1971)
	Wroclaw, Poland	Carp ponds	Carp ponds	F	*Cyprinus carpio*	*Cyprinus carpio* Linnaeus, 1758	Wagler (1935) as cited in Grabda (1971)
	South East England	Koi ponds	Koi ponds	F	Koi carp	*Cyprinus carpio* Linnaeus, 1758	Gresty et al. (1993)
Argulus matritensis Arévalo, 1921[f]	Seville, Spain	Guadalquivir River	Guadalquivir River	F	Free swimming (Plankton)	Free swimming (Plankton)	Arévalo (1921)
	Madrid, Spain	pond in Retiro	pond in Retiro	F	Free swimming (Plankton)	Free swimming (Plankton)	
Argulus pellucidus Wagler, 1935[g]	Bavaria, Germany	Bavarian Biological Research Institute for Fisheries in Wielenbach	Bavarian Biological Research Institute for Fisheries in Wielenbach	F	Not given	Not given	Wagler (1935)
Argulus phoxini Leydig, 1871	Bach, Germany	Goldersbach	Goldersbach creek	F	*Phoxinus laevis*	*Phoxinus phoxinus* (Linnaeus, 1758)	Leydig (1871)
	Tübingen			F	*Phoxinus laevis*	*Phoxinus phoxinus* (Linnaeus, 1758)	Wilson (1902)

TABLE III
(Continued)

Argulus species	Country/city/town	Water system as published	Water system confirmed[a]	E	Host species	Host species confirmed[b]	Reference
Argulus purpureus Risso, 1826[h]	Mediterranean	Not given	Not given	U	Not given	Not given	Risso (1826)
Argulus rothschildi Leigh-Sharpe, 1933[i]	Hertfordshire, England	Tring Reservoir	Tring Reservoirs	F	*Abramis brama*	*Abramis brama* (Linnaeus, 1758)	Leigh-Sharpe (1933)
Argulus viridis Nettovich, 1900[j]	Not given	Not given	Not given	F	*Rhodeus amarus*	*Rhodeus amarus* (Bloch, 1782)	Nettovich (1900)
				F	*Phoxinus laevis*	*Phoxinus phoxinus* (Linnaeus, 1758)	
				F	*Alburnus lucidus*	*Alburnus alburnus* (Linnaeus, 1758)	

E, Environment in which it is found: B, brackish water; F, freshwater; M, marine; U, unknown.

[a]) Water system confirmed.

[b]) Host species confirmed using fishbase.org.

[c]) First described from North America.

[d]) *A. giordanii* Brian, 1959 is listed by WoRMS but while the reference is known, it was not located.

[e]) Cosmopolitan distribution.

[f]) *A. matritensis* Arévalo, 1921 synonym of *A. japonicus* Thiele, 1900 accordings to Fryer (1982).

[g]) *A. pellucidus* Wagler, 1935 synonym of *A. japonicus* Thiele, 1900 according to Fryer (1960a).

[h]) *A. purpureus* Risso, 1826 synonym of *A. vittatus* (Rafinesque-Schmaltz, 1814) according to Thorell (1864).

[i]) *A. rothschildi* Leigh-Sharpe, 1933 synonym of *A. foliaceus* (Linnaeus, 1758) according to Gurney (1948).

[j]) *A. viridis* Nettovich, 1900 synonym of *A. foliaceus* (Linnaeus, 1758) according to Romanovsky (1955).

TABLE IV
Marine, brackish water and Freshwater *Argulus* spp. recorded from Asia

Argulus species	Country/city/town	Water system as published	Water system confirmed[a]	E	Host species	Host species confirmed[b]	Reference
Argulus belones van Kampen, 1909	Olehleh, Sumatra	North Coast	Andaman Sea, Indian Ocean	M	*Belone schismatorhynchus* Blkr.	*Ablennes hians* (Valenciennes, 1846)	Van Kampen (1909)
Argulus bengalensis Ramakrishna, 1951	Harischandrapur, Malda District, West Bengal, India	Unknown	Unknown	U	Host unknown	Host unknown	Ramakrishna (1951)
	Guskara Guskara (23°30′N 87°45′E), Churulia (23°47′N, 87°0′E) and Kalna (23°13′N 88°22′E), West Bengal, India	commercial fish farms	commercial fish farms	F	*Cirrhinus mrigala* (Hamilton, 1822)	*Cirrhinus mrigala* (Hamilton, 1822)	Guha et al. (2012)
	Guskara (23°50′N 87°45′E) Burdwan, India	commercial fish farms	commercial fish farms	F	*Cirrhinus mrigala* (Hamilton, 1822)	*Cirrhinus mrigala* (Hamilton, 1822)	Guha et al. (2013)
	Malda, West Bengal, India	Barasagar Dighi fish farm 24°58′08.86″N 88°06′09.70″E	Barasagar Dighi fish farm 24°58′08.86″N 88°06′09.70″E	F	*Cirrhinus mrigala* (Hamilton, 1822)	*Cirrhinus mrigala* (Hamilton, 1822)	Banerjee & Saha (2013, 2016); Banerjee et al. (2014a, b, 2015)
	West Bengal, India	University laboratory culture	University laboratory culture	F	Not given	Not given	Banerjee et al. (2014b)

TABLE IV
(Continued)

Argulus species	Country/city/town	Water system as published	Water system confirmed[a]	E	Host species	E	Host species confirmed[b]	Reference
Argulus boli Tripathi, 1975[c]								
Argulus caecus Wilson, 1922	Sagami, Japan	Aburatsubo	Aburatsubo, Pacific Ocean	M	Host unknown	M	Host unknown	Wilson (1922)
	Japan	Pacific Coast	Pacific Coast	M	Spheroides spp.	M	Spheroides spp.	Tokioka (1936a)
Argulus cauveriensis Thomas & Deveraj, 1975	Hogenakkal, India	Kaveri River	Kaveri River	F	Free swimming	F	Free swimming	Thomas & Deveraj (1975)
Argulus cheni Shen, (1948	Canton, P. R. China	Unknown	Unknown	U	Not given	U	Not given	Shen (1948)
Argulus chinensis Ku & Wang, 1955	Soochow, P. R. China		Suzhou, Jiangsu, P. R. China	B	Opiocephalus argus	B	Channa argus (Cantor, 1842)	Wang (1958)
	Wushi, P. R. China		Wushi, Xiangtan	B	Mylopharyngodon aethiops	B	Mylopharyngodon piceus (Richardson, 1846)	Wang (1958)
				B	Leiocassis sp.	B	Leiocassis sp.	
Argulus coregoni Thorell, 1866[d]	Ôtsu, Japan	Unknown	Unknown	U	Acheilognathus moriokae Jordan & Thompson	U	Acheilognathus melanogaster Bleeker, 1860	Tokioka (1936a)
	Nagano, Tokyo	fish culture ponds	fish culture ponds	F	Salvelinus frontalis	F	Ecsenius frontalis (Valenciennes, 1836)	Hoshina (1950)
	Nagano, Tokyo	fish culture ponds	fish culture ponds	F	Salmo irideus	F	Oncorhynchus mykiss (Walbaum, 1792)	

TABLE IV
(Continued)

Argulus species	Country/city/town	Water system as published	Water system confirmed[a]	E	Host species		Host species confirmed[b]	Reference
	Nagano, Tokyo	fish culture ponds	fish culture ponds	F	Plecoglossus altivelis		Plecoglossus altivelis altivelis (Temminck & Schlegel, 1846)	Shimura (1981, 1983a); Shimura & Inoue (1984)
	Tokyo, Japan	Okutama fisheries experiment station 35°49'N 139°09'E	Okutama fisheries experiment station 35°49'N 139°09'E	F	Salmos gairdneri		Oncorhynchus mykiss (Walbaum, 1792)	Shimura (1981, 1983a); Shimura & Inoue (1984)
	Tokyo, Japan	Okutama fisheries experiment station 35°49'N 139°09'E	Okutama fisheries experiment station 35°49'N 139°09'E	F	Oncorhynchus maso		Oncorhynchus masou masou (Brevoort, 1856)	Shimura (1981, 1983a, b); Shimura & Inoue (1984)
	Soochow, P. R. China	Suzhou, Jiangsu, P. R. China		B	Mylopharyngodon aethiops		Mylopharyngodon piceus (Richardson, 1846)	
	Kinki University, Shingu City, Japan	fisheries laboratory	fisheries laboratory	F	Oncorhynchus masou ishikawai		Oncorhynchus masou masou (Brevoort, 1856)	Nagasawa & Ohya (1996a)
	Kinki University, Shingu City, Japan	fisheries laboratory	fisheries laboratory	F	Plecoglossus altivelis		Plecoglossus altivelis altivelis (Temminck & Schlegel, 1846)	Nagasawa & Ohya (1996b)
	Yoshiga, Shimane Prefecture, Japan	Tachigochi River 34°20'08"N 131°57'48"E	Tachigochi River 34°20'08"N 131°57'48"E	F	Salvenus leucomaenis imbrius gogi charr		Salvelinus leucomaenis imbrius Jordan & McGregor (1925)	Nagasawa & Kawai (2008)

TABLE IV
(Continued)

Argulus species	Country/city/town	Water system as published	Water system confirmed[a]	E	Host species	Host species confirmed[b]	Reference
	Selangor, Malaysia	Langat River	Langat River	F	Red tilapia	Red tilapia	Everts & Avenant-Oldewage (2009)
	Kinki University, Shingu City, Japan	fisheries laboratory	fisheries laboratory	F	*Oncorhynchus masou ishikawai*	*Oncorhynchus masou masou* (Brevoort, 1856)	Kaji et al. (2011)
Argulus ellipticaudatus Wang, 1960	Soochow, P. R. China		Suzhou, Jiangsu, P. R. China	B	*Cyprinus carpio*	*Cyprinus carpio* Linnaeus, 1758	Wang (1960)
Argulus foliaceus Linnaeus, 1758[d]	Canton, P. R. China	fisheries experiment station	fisheries experiment station	F	*Ctenopharyngodon idellus* (Cuv. & Val.)	*Ctenopharyngodon idella* (Valenciennes, 1844)	Chen (1933)
	Yin-On Shah, P. R. China	Not given	Not given	U	*Carrasius auratus*	*Carassius auratus* (Linnaeus, 1758)	Wang (1958)
	Abbey park, Leuven	breeding ponds	breeding ponds	F	*Leuciscus rutilus* L.	*Rutilus rutilus* (Linnaeus, 1758)	Van den Bosch de Aguilar (1972)
	Baghdad	Al-Nibaey Lake-two	Al-Nibaey Lake-two	F	*Carasobarbus luteus*	*Carasobarbus luteus* (Heckel, 1843)	Ali et al. (1988)
	Baghdad	Al-Tharthar Canal	Al-Tharthar Lake Canal	F	*Barbus xanthopterus*	*Luciobarbus xanthopterus* Heckel, 1843	Khalifa (1989)
				F	*Cyprinus carpio*	*Cyprinus carpio* Linnaeus, 1758	

TABLE IV
(Continued)

Argulus species	Country/city/town	Water system as published	Water system confirmed[a]	E	Host species	Host species confirmed[b]	Reference
	Turkey	Seyhan River	Seyhan River	F	Cyprinus carpio L., 1758	Cyprinus carpio Linnaeus, 1758	Cengizler et al. (2001)
	Turkey	pet store	pet store	F	Carassius auratus	Carassius auratus (Linnaeus, 1758)	Yildiz & Kumantas (2002)
	Sri Lanka	ornamental fish farm	ornamental fish farm	F	Carassius auratus goldfish	Carassius auratus (Linnaeus, 1758)	Thilakaratne et al. (2003)
				F	Xiphophorus maculatus platy	Xiphophorus maculatus (Günther, 1866)	
				F	Cyprinus carpio Carp	Cyprinus carpio Linnaeus, 1758	
	Turkey	Kovada Lake	Kovada Lake	F	Carassius auratus	Carassius auratus (Linnaeus, 1758)	Özan & Kir (2005)
	Afyon, Turkey	Lake Eber	Lake Eber	F	Cyprinus carpio L.	Cyprinus carpio Linnaeus, 1758	Öztürk (2005)
	Turkey	Kovada Lake	Kovada Lake	F	Carassius carassius L., 1758	Carassius carassius (Linnaeus, 1758)	Tekin Özan & Kir (2005)
	Turkey	Lake Durusu (Terkos)	Terkos Dam	F	Abramis brama	Abramis brama (Linnaeus, 1758)	Karatoy & Soylu (2006)
	Turkey	Atatürk Dam Lake	Atatürk Dam	F	Liza abu Heckel, 1843	Liza abu (Heckel, 1843)	Öktener et al. (2006)
				F	Mastacembelus mastacembelus L., 1758 spiny eel	Mastacembelus mastacembelus (Banks & Solander, 1794)	

TABLE IV
(Continued)

Argulus species	Country/city/town	Water system as published	Water system confirmed[a]	E	Host species	Host species confirmed[b]	Reference
				F	Silurus triostegus L., 1758 Asian catfish	Silurus triostegus Heckel, 1843	
	Turkey	The Sapanca Lake	Lake Sapanca	F	Vimba vimba Linnaeus, 1758	Vimba vimba (Linnaeus, 1758)	Uzunay & Soylu (2006)
	Marivan city, Kurdistan Province, Iran	Zarivar Lake	Lake Zarivar	F	Hypophthalmichthys molitrix Silver carp	Hypophthalmichthys molitrix (Valenciennes, 1844)	Jalali & Barzegar (2006)
	Marivan city, Kurdistan Province, Iran	Zarivar Lake	Lake Zarivar	F	Cyprinus carpio Linnaeus, 1758 common carp	Cyprinus carpio Linnaeus, 1758	
	Marivan city, Kurdistan Province, Iran	Zarivar Lake	Lake Zarivar	F	Chalcalburnus sp. Bleak	Alburnus sp.	
	Marivan city, Kurdistan Province, Iran	Zarivar Lake	Lake Zarivar	F	Mastacembelus mastacembelus spiny eel	Mastacembelus mastacembelus (Banks & Solander, 1794)	Jalali & Barzegar (2006); Jalali et al. (2008)
	Konya, Turkey	Çavuşçu Lake	Çavuşçu Lake	F	Cyprinus carpio Lin., 1758	Cyprinus carpio Linnaeus, 1758	Öktener et al. (2007)
	Konya, Central Anatolia, Turkey	Çavuşçu Lake 36°41'N 34°26'E	Çavuşçu Lake 36°41'N 34°26'E	F	Cyprinus carpio Linnaeus, 1758	Cyprinus carpio Linnaeus, 1758	Alaş et al. (2010)
	Esfahan, Iran	goldfish producer	goldfish producer	F	Carassius auratus	Carassius auratus (Linnaeus, 1758)	Noaman et al. (2010)
	Turkey	Enne Dam Lake	Enne Dam	F	Alburnus alburnus	Alburnus alburnus (Linnaeus, 1758)	Koyun (2011)

TABLE IV
(Continued)

Argulus species	Country/city/town	Water system as published	Water system confirmed[a]	E	Host species	Host species confirmed[b]	Reference
Argulus fluviatilis Thomas & Deveraj, 1975[e]	Samsun, Turkey	Fish farm	Fish farm	F	Carassius carassius	Carassius carassius (Linnaeus, 1758)	Pekmezci et al. (2011)
					Carassius auratus	Carassius auratus (Linnaeus, 1758)	
	Hogenakkal, India	Kaveri River	Kaveri River	F	Cyprinus carpio	Cyprinus carpio Linnaeus, 1758	Thomas & Deveraj (1975)
				F	Free swimming	Free swimming	
Argulus giganteus Ramakrishna, 1951	Unknown	Unknown	Unknown	U	Host unknown	Host unknown	Ramakrishna (1951)
Argulus indicus Weber, 1892	Bombay, India	Mahim Creek	Mahim Creek	F	Tetrodon oblongus	Takifugu oblongus (Bloch, 1786)	Rangnekar (1957)
	Sumatra	East Indian Archipelago	East Indian Archipelago	F	Not given	Not given	Weber (1892)
	Not given	East Indian Archipelago	East Indian Archipelago	B	Host unknown	Host unknown	Wilson (1902)
	Thailand (published as Siam)	Unknown	Unknown	U	Fighting fish	Betta splendens Regan (1910	Wilson (1927)
	Bangkok		Gulf of Thailand	M	Trichopodus pectoralis	Trichopodus pectoralis Regan (1910	Wilson (1944)
	Champahati, West Bengal, India	Unknown	Unknown	U	Ophiocephalus punctatus	Channa punctata (Bloch, 1793)	Ramakrishna (1951)
	Perak	Tasik Temengor	Temenggor Lake	F	Ophiocephalus micropeltes	Channa micropeltes (Cuvier, 1831)	Seng (1986)

TABLE IV
(Continued)

Argulus species	Country/city/town	Water system as published	Water system confirmed[a]	E	Host species	Host species confirmed[b]	Reference
	Jyotisar, India	Haryana Government Fish Seed Farm	Haryana Government Fish Seed Farm	E			Singhal et al. (1986)
				F	*Catla catla*	*Catla catla* (Hamilton, 1822)	
				F	*Labeo rohita*	*Labeo rohita* (Hamilton, 1822)	
				F	*Cirrhinus mrigala*	*Cirrhinus mrigala* (Hamilton, 1822)	
				F	*Ctenopharyngodon idella*	*Ctenopharyngodon idella* (Valenciennes, 1844)	
				F	*Hypophthalmichthys molitrix*	*Hypophthalmichthys molitrix* (Valenciennes, 1844)	
	Karnal, India	Central Inland Fisheries Research Institute	Central Inland Fisheries Research Institute	F	*Catla catla*	*Catla catla* (Hamilton, 1822)	
				F	*Labeo rohita*	*Labeo rohita* (Hamilton, 1822)	
				F	*Cirrhinus mrigala*	*Cirrhinus mrigala* (Hamilton, 1822)	
				F	*Ctenopharyngodon idella*	*Ctenopharyngodon idella* (Valenciennes, 1844)	

TABLE IV
(Continued)

Argulus species	Country/city/town	Water system as published	Water system confirmed[a]	E	Host species	Host species confirmed[b]	Reference
				F	*Hypophthalmichthys molitrix*	*Hypophthalmichthys molitrix* (Valenciennes, 1844)	
	Jyotisar, India	Haryana Government Fish Seed Farm	Haryana Government Fish Seed Farm	F	*Cyprinus carpio*	*Cyprinus carpio* Linnaeus, 1758	Singhal et al. (1990)
	Karnal, India	Central Inland Fisheries Research Institute	Central Inland Fisheries Research Institute	F	*Cyprinus carpio*	*Cyprinus carpio* Linnaeus, 1758	
	Tando Allahyar, Hydrabad district, Sindh, Pakistan	fish pond	fish pond	F	Major carps	*Cyprinus carpio* Linnaeus, 1758	Jafri & Ahmed (1991)
Argulus japonicus Thiele, 1900	Yeddo, Japan	Not given	Not given	U	Gotsche	Gotsche	Thiele (1900)
	Japan	Unknown	Unknown	F	Goldfish	*Carassius auratus* (Linnaeus, 1758)	Tokioka (1936a)
				F	*Cyprinus carpio* (Linné)	*Cyprinus carpio* Linnaeus, 1758	
				F	*Carassius carassius* (Linné)	*Carassius carassius* (Linnaeus, 1758)	
	Kôriyama, Japan	Unknown	Unknown	U	Host unknown	Host unknown	Tokioka (1936b)
	Mishan, Manchukuo	Not given	Not given	U	Host unknown	Host unknown	Tokioka (1939)

TABLE IV
(Continued)

Argulus species	Country/city/town	Water system as published	Water system confirmed[a]	E	Host species	E	Host species confirmed[b]	Reference
	Yaizu, Japan	aquarium of Yaizu prefectural school of fishery	aquarium of Yaizu prefectural school of fishery		Goldfish	F	*Carassius auratus* (Linnaeus, 1758)	Yamaguti (1937)
	Yunnan, P. R. China	Lake Erh Hai (Tali Lake)	Erhai Lake		Plankton	F	Free swimming	Hsiao (1950)
	Tokyo, Japan	Ueno Zoo Aquarium	Ueno Zoo Aquarium		*Carassius carassius auratus*	F	*Carassius auratus* (Linnaeus, 1758)	Shimura (1983b)
	Batu Berendam, Malacca, Malaysia	Malacca River	Malacca River		*Ctenopharyngodon idellus*	F	*Ctenopharyngodon idella* (Valenciennes, 1844)	Seng (1986)
	Japan	Lake Onuma	Lake Onuma		*Cyprinus carpio*	F	*Cyprinus carpio* Linnaeus, 1758	Nagasawa et al. (1989)
	Hyderabad and Thatta, Sindh, Pakistan	fish farms	fish farms		*Labeo rohita*	F	*Labeo rohita* (Hamilton, 1822)	Jafri & Ahmed (1991)
					Catla catla	F	*Catla catla* (Hamilton, 1822)	
					Cirrhina mrigala	F	*Cirrhinus mrigala* (Hamilton, 1822)	
	Ibaraki, Japan	fisheries station	fisheries station		Mirror carps	F	*Cyprinus carpio* Linnaeus, 1758	Ikuta & Makioka (1994, 1997); Ikuta et al. (1997)
	Chonbuk, South Korea	Okjeong lake	Lake Ok-jeong		*Parasilurus asotus* (cat fish)	F	*Silurus asotus* Linnaeus, 1758	Han et al. (1998)

TABLE IV
(Continued)

Argulus species	Country/city/town	Water system as published	Water system confirmed[a]	E	Host species	Host species confirmed[b]	Reference
				F	*Ictalurus punctatus* (channel catfish)	*Ictalurus punctatus* (Rafinesque, 1818)	
				F	*Carassius carassius* (Crusian carp)	*Carassius carassius* (Linnaeus, 1758)	
				F	*Cyprinus carpio* (Israeli & Korean carp)	*Cyprinus carpio* Linnaeus, 1758	
	Shunde city, Guangdong Province, P. R. China	fish farms	fish farms	F	Host unknown	Host unknown	Wadeh et al. (2008)
	Kanagawa, Japan	pond (35°22′N, 139°28′E)	pond (35°22′N, 139°28′E)	F	*Cyprinus carpio* carp	*Cyprinus carpio* Linnaeus, 1758	Yoshizawa & Nogami (2008)
	Dehar, Sundernager, Mandi, Himachal Pradesh, India	fish farms	fish farms	F	*Cyprinus carpio* carp	*Cyprinus carpio* Linnaeus, 1758	Sahoo et al. (2012)
	Guangdong Province, P. R. China (20°13′N/109°39′ to 117°19′E)	various locations	various locations	U	Not given	Not given	Alsarakibi et al. (2014)
Argulus kunmingensis Shen, 1948	Trimonument Market, Kunming, Yunnan, P. R. China	fish basin	fish basin	U	Free swimming	Free swimming	Shen (1948)
	Kwan Ying Shan	Kunming Lake	Kunming Lake	F	Free swimming	Free swimming	

TABLE IV
(Continued)

Argulus species	Country/city/town	Water system as published	Water system confirmed[a]	E	Host species	Host species confirmed[b]	Reference
Argulus kusafugu Yamaguti & Yamasu, 1959	Shinmaiko, Japan		Ise Bay	M	Spheroides niphobles	Takifugu niphobles (Jordan & Snyder, 1901)	Yamaguti & Yamasu (1959)
Argulus major Wang, 1960	Ling-Ho Chee-Kiang		Unknown	U	Host unknown	Host unknown	Wang (1960)
Argulus mangalorensis Natarajan, 1982	Mangalore, India	Nethravarthy	Netravati River	F	Free swimming	Free swimming	Natarajan (1982)
Argulus matuii Sikama, 1938	Tiba Prefecture, Japan		Pacific Ocean	M	Parapristipoma trilineatum	Parapristipoma trilineatum (Thunberg, 1793)	Sikama (1938)
Argulus melanostictus Wilson, 1935[f]	Thailand	Gulf of Thailand	Gulf of Thailand	M	Free swimming	Free swimming	Wilson (1944)
Argulus mongolianus Tokioka, 1939	Manchukuo	Dalai-nor	Lake Hulun	F	Host unknown	Host unknown	Tokioka (1939)
Argulus nativus Kirtisinghe, 1959	Ceylon	South West Coast	Indian ocean	M	Promicrops lanceolatus (Bloch)	Epinephelus lanceolatus (Bloch, 1790)	Kirtisinghe (1959)
	Ambalanga, Ceylon (Sri Lanka)	South West Coast	Indian ocean	M	Promicrops lanceolatus (Bloch)	Epinephelus lanceolatus (Bloch, 1790)	Kirtisinghe (1964)
Argulus onodai Tokioka, 1936	Wakayama, Japan	Katu-ura Kii	Kii-Katuura, North Pacific Ocean	M	Spheroides alboplumbeus (Richardson)	Takifugu alboplumbeus (Richardson, 1845)	Tokioka (1936a)
Argulus parsi Tripathi, 1975[c]							

TABLE IV
(Continued)

Argulus species	Country/city/town	Water system as published	Water system confirmed[a]	E	Host species	Host species confirmed[b]	Reference
Argulus plecoglossi Yamaguti, 1937	Japan	River Hozu	Hozu River	F	Plecoglossus altivelis Temm. & Schleg.	Plecoglossus altivelis altivelis (Temminck & Schlegel, 1846)	Yamaguti (1937)
Argulus puthenveliensis Ramakrishna, 1959[c]	Ernakulam, Kerala State, India	Udayamperoor pond	Valiyakulam Pond	F	Esomus danrica (Ham.)	Esomus danricus (Hamilton, 1822)	Thomas (1961)
Argulus quadristriatus Deveraj & Ameer Hamsa, 1977	Mandapam, India	Palk Bay	Palk Strait	F	Psammoperca waigiensis (Cuvier)	Psammoperca waigiensis (Cuvier, 1828)	Deveraj & Ameer Hamsa (1977)
Argulus scutiformis Thiele, 1900	Japan	Not given	Not given	U	Not given	Not given	Thiele (1900)
	Misaki, Japan	Unknown	Unknown	M	Mola mola (Linné)	Mola mola (Linnaeus, 1758)	Tokioka (1936a)
	Hokkaido, Japan	Unknown	Unknown	U	Host unknown	Host unknown	
	Shinmaiko, Japan	Not given	Ise Bay	M	Spheroides rubriceps (Temm. & Schleg.)	Takifugu sp.	Yamaguti & Yamasu (1959)
Argulus siamensis Wilson, 1926	Bangkok	private pond	private pond	F	Cirrhina	Cirrhina sp.	Wilson (1926)
	Harischandrapur, Malda District, West Bengal, India	Unknown	Unknown	U	Host unknown	Host unknown	Ramakrishna (1951)
	Champahati, Calcutta, West Bengal	Unknown	Unknown	U	Ophicephalus punctatus Bloch	Channa punctata (Bloch, 1793)	

TABLE IV
(Continued)

Argulus species	Country/city/town	Water system as published	Water system confirmed[a]	E	Host species	Host species confirmed[b]	Reference
	Siripur, Bihar (specimen of Dr. T. Southwell, recorded as A. foliaceus)	Unknown	Unknown	U	Labeo rohita	Labeo rohita (Hamilton, 1822)	
	Siliguri, base of Himalayas	Mahananda river	Mahananda river	F	Host unknown	Host unknown	
	Dharangadhara State (Saurashtra) (specimen of Dr S. L. Hora, recorded as A. foliaceus)	Unknown	Unknown	U	Murrel	Channa striata (Bloch, 1793)	
	Chitradurga district, Karnataka, India	Vani Vilas Sagar fish farm	Vani Vilas Sagar fish farm	F	Catla catla	Catla catla (Hamilton, 1822)	Sundara Bai et al. (1988)
				F	Ctenopharyngodon idella	Ctenopharyngodon idella (Valenciennes, 1844)	
	Bangalore, India	fish farm at Hessaraghatta	fish farm at Hessaraghatta	F	Lebistes reticulatus	Poecilia reticulata Peters, 1859	Nandp & Das (1991)
	Kakdwip, West Bengal, India	fish farm at Kakdwip	fish farm at Kakdwip,	F	Labeo rohita	Labeo rohita (Hamilton, 1822)	
					Cirrhinus mrigala (Hamilton)	Cirrhinus mrigala (Hamilton, 1822)	

TABLE IV
(Continued)

Argulus species	Country/city/town	Water system as published	Water system confirmed[a]	E	Host species	Host species confirmed[b]	Reference
	Ramsager, Bankura District, West Bengal, India	freshwater fish farm 23°05'59.66"N 87°16'27.44"E	freshwater fish farm 23°05'59.66"N 87°16'27.44"E	F	Calla catla (Hamilton) / Cirrhinus mrigala (Hamilton, 1822)	Calla catla (Hamilton, 1822) / Cirrhinus mrigala (Hamilton, 1822)	Saha et al. (2011)
	Bangalore, India	Regional Research Centre of C.I.F.A. culture pond	Regional Research Centre of C.I.F.A. culture pond	F	Labeo rohita	Labeo rohita (Hamilton, 1822)	Hemaprasanth et al. (2012); Sahoo et al. (2013)
Argulus siamensis-peninsularis Ramakrishna, 1951	Rajahmundry	Unknown	Unknown	U	Host unknown	Host unknown	Ramakrishna (1951)
	Jabalpur, India	Gokalpur Lake	Gokalpur Lake	F	Ambassis ranga	Parambassis ranga (Hamilton, 1822)	Malaviya (1955)
Argulus sindhensis Mahar & Jafri, 2011	Thatta District, Pakistan	fish farm	fish farm	F	Labeo rohita	Labeo rohita (Hamilton, 1822)	Mahar & Jafri (2011)
Argulus taliensis Shen, 1948	Kiang-Shian-Tsun, Yunnan, P. R. China	Er Hai, Lake of Tali	Erhai Lake	F	Free swimming	Free swimming	Shen (1948)
Argulus tientsinensis Ku & Wang, 1956	Tientsin, P. R. China	Not given	Not given	U	Pseudobargrus fulvidraco (Richardson) yellow-barbed catfish	Tachysurus fulvidraco (Richardson, 1846)	Ku & Wang (1956)
Argulus trilineatus Wilson, 1904[f]	Yunnan, P. R. China	Lake Erh Hai (Tali Lake)	Erhai Lake	F	Plankton	Free swimming	Hsiao (1950)

TABLE IV
(Continued)

Argulus species	Country/city/town	Water system as published	Water system confirmed[a]	E	Host species	Host species confirmed[b]	Reference
Argulus tristramellae Paperna, 1964	Israel	Lake Tiberias	Sea of Galilee	M	Tristramella simonis (Gunther)	Tristramella simonis simonis (Günther, 1864)	Paperna (1964)
	Israel	Lake Kinneret	Sea of Galilee	M	Tristramella simonis (Gunther)	Tristramella simonis simonis (Günther, 1864)	Landsberg (1989)
Argulus yuii Wang, 1958	Soochow, P. R. China		Suzhou, Jiangsu, P. R. China	B	Cyprinus carpio	Cyprinus carpio Linnaeus, 1758	Wang (1958)
				B	Mylopharyngodon aethiops	Mylopharyngodon piceus (Richardson, 1846)	
Argulus yunnanensis Shen, 1948	Kunming, P. R. China	Kunming Lake	Kunming Lake	F	Free swimming	Free swimming	Shen (1948)

E, environment in which it is found: B, brackish water; F, freshwater; M, marine; U, unknown.
[a] Water system confirmed.
[b] Host species confirmed using fishbase.org.
[c] A. boli Tripathi (1975); A. parsi Tripathi (1975); A. puthenveliensis Ramakrishna (1959) are listed by WoRMS but the references while known, were not found.
[d] First described from Europe.
[e] This is a homonym that needs to be renamed because it is already in use A. giganteus Lucas, 1849 according to Rushton-Mellor (1594b).
[f] First described from North America.

TABLE V
Marine, brackish water and freshwater *Argulus* spp. recorded from Africa

Argulus species	Country/city/town	Water system as published	Water system confirmed[a]	Host species	E	Host species confirmed[b]	Reference
Argulus africanus Thiele, 1900	Langenburg (Tukuyu, Tanzania)		Lake Nyassa	*Clarias* sp.	F	*Clarias* sp.	Thiele (1900)
	Kirima, Tanzania	Lake Albert-Edward	Lake Albert-Edward	Free swimming	F	Free swimming	
	Not given	Nile river	Nile River	Free swimming	F	Free swimming	Cunnington (1913)
	Sumbu	Lake Tanganyika	Lake Tanganyika	*Clarias robecchii*	F	*Clarias gariepinus* (Burchell, 1822)	
	Vua	Lake Tanganyika	Lake Tanganyika	*Clarias lazera* Muomi	F	*Clarias gariepinus* (Burchell, 1822)	
	Kala	Lake Tanganyika	Lake Tanganyika	*Lates microlepis*	F	*Lates microlepis* Boulenger, 1898	
	Ndanvie	Lake Tanganyika	Lake Tanganyika	*Lates microlepis*	F	*Lates microlepis* Boulenger, 1898	
	Bukoba	Lake Victoria, Nyanza	Lake Victoria	*Bagrus degeni* Nfui	F	*Bagrus degeni* Boulenger (1906)	
	Bukoba	Lake Victoria, Nyanza	Lake Victoria	*Protopterus ethiopicus*	F	*Protopterus aethiopicus* Heckel, 1851	
	Bukoba	Lake Victoria, Nyanza	Lake Victoria	*Clarias anguillaris* Nshonzi	F	*Clarias anguillaris* (Linnaeus, 1758)	
	Mugungo	Lake Albert Nyanza	Lake Albert	Free swimming	F	Free swimming	
	Not given	Lake No, White Nile	Lake No, White Nile	*Heterobranchus bidorsalis*	F	*Heterobranchus bidorsalis* Geoffroy Saint-Hilaire, 1809	

TABLE V
(Continued)

Argulus species	Country/city/town	Water system as published	Water system confirmed[a]	E	Host species	Host species confirmed[b]	Reference
	Not given	Lake Nyasa	Lake Malawi	F	*Bagrus meridionalis* Günther	*Bagrus meridionalis* Günther, 1894	Fryer (1956)
				F	*Anguilla nebulosa labiata* Peters	*Anguilla labiata* (Peters, 1852)	
				F	*Mormyrus longirostris* Peters	*Mormyrus longirostris* Peters, 1852	
	Zimbabwe	Lake Bangweulu	Lake Bangweulu	F	*Heterobranchus longifilis* Cuvier & Valenciennes	*Heterobranchus longifilis* Valenciennes, 1840	Fryer (1959)
	Zimbabwe	Mulungushi River	Mulungushi River	F	*Heterobranchus* sp.	*Heterobranchus* sp.	Fryer (1960a)
	Not given	Lake Mweru	Lake Mweru	F	*Mormyrops deliciosus* (Leach)	*Mormyrops anguilloides* (Linnaeus, 1758)	
	Not given	Lake Mweru	Lake Mweru	F	*Heterobranchus longifilis* Cuvier & Valenciennes	*Heterobranchus longifilis* Valenciennes, 1840	
	Not given	Lake Mweru	Lake Mweru	F	*Eutropius banguelensis* Boulenger	*Schilbe banguelensis* (Boulenger, 1911)	
	Not given	Lake Victoria	Lake Victoria	F	*Protopterus aethiopicus* Heckel	*Protopterus aethiopicus* Heckel, 1851	Fryer (1961a)

TABLE V
(Continued)

Argulus species	Country/city/town	Water system as published	Water system confirmed[a]	E	Host species	Host species confirmed[b]	Reference
				F	*Clarias mossambicus* Peters	*Clarias gariepinus* (Burchell, 1822)	
				F	*Bagrus doemac* Forskal	*Bagrus docmak* (Forsskål, 1775)	
				F	*Tilapia variabilis* Boulenger	*Oreochromis variabilis* (Boulenger, 1906)	
				F	*Tilapia esculenta* Graham	*Orecchromis esculentus* (Graham, 1928)	
	Not given	Lake Victoria	Lake Victoria	F	*Tilapia variabilis* Boulenger	*Oreochromis variabilis* (Boulenger, 1906)	Fryer (1963)
	Not given	confluence of Matetsi and Deka Rivers with Zambezi 18°05′S 26°40′E	confluence of Matetsi and Deka Rivers with Zambezi 18°05′S 26°40′E	F	*Clarias gariepinus* Burchell	*Clarias gariepinus* (Burchell, 1822)	Fryer (1964)
	Not given	Lake Mweru	Lake Mweru	F	*Tilapia macrochir* Boulenger	*Oreochromis macrochir* (Boulenger, 1912)	Fryer (1965a)
	Not given	Lake Edward	Lake Edward	F	*Clarias* sp.	*Clarias* sp.	
	Not given	Lake Tanganyika	Lake Tanganyika	F	*Heterobranchus* sp.	*Heterobranchus* sp.	

TABLE V
(Continued)

Argulus species	Country/city/town	Water system as published	Water system confirmed[a]	E	Host species	Host species confirmed[b]	Reference
	Not given	Lake Tanganyika	Lake Tanganyika	F	*Hydrocyon lineatus* Bleeker	*Hydrocynus vittatus* Castelnau, 1861	
	Not given	Lake Tanganyika	Lake Tanganyika	F	*Chrysichthys brachynema* Boulenger	*Chrysichthys brachynema* Boulenger (1900)	
	Not given	Lake Tanganyika	Lake Tanganyika	F	*Polypterus* sp.	*Polypterus* sp.	Fryer (1968)
	Tanzania	Lake Kitangiri	Lake Kitangiri	F	*Protopterus*	*Protopterus* sp.	
				F	*Clarias*	*Clarias* sp.	
				F	*Schilbe*	*Schilbe* sp.	
				F	*Labeo*	*Labeo* sp.	
				F	*Tilapia*	*Tilapia* sp.	
	Uganda	Lake Victoria	Lake Victoria	F	*Hoplotilapia retrodens*	*Haplochromis retrodens* (Hilgendorf, 1888)	Thurston (1970)
	Uganda	Lake Victoria	Lake Victoria	F	*Haplochromis guiarti*	*Haplochromis guiarti* (Pellegrin, 1904)	
	Uganda	Lake Victoria	Lake Victoria	F	*Haplochromis obesus*	*Haplochromis obesus* (Boulenger, 1906)	
	Uganda	Lake Victoria	Lake Victoria	F	*Haplochromis obliquidens*	*Haplochromis obliquidens* (Hilgendorf, 1888)	
	Uganda	Lake Victoria	Lake Victoria	F	*Bagrus docmac* Forskahl	*Bagrus docmak* (Forsskål, 1775)	Mbahinzireki (1980)

TABLE V
(Continued)

Argulus species	Country/city/town	Water system as published	Water system confirmed[a]	E	Host species	Host species confirmed[b]	Reference
Argulus alexandrensis Wilson, 1923	Nigeria	Lake Kainji	Kainji Lake	F	*Oreochromis niloticus*	*Oreochromis niloticus* (Linnaeus, 1758)	Okaeme et al. (1988)
	Porto Alexandre, Tombua, Angola	Port Alexandre	South Atlantic Ocean	M	*Zeus* sp.	*Zeus* sp.	Wilson (1923)
Argulus ambloplites Wilson, 1920[c]	Faradje, Congo	Dungu River	Dungu River	F	*Ophiocephalus obscurus* Günther	*Parachanna obscura* (Günther, 1861)	Wilson (1920b)
	Zimbabwe	Lake Bangweulu	Lake Bangweulu	F	*Clarias mossambicus* Peters	*Clarias gariepinus* (Burchell, 1822)	Fryer (1959)
	Tanzania	Lake Chaya	Lake Chaya	F	*Hydrocyon lineatus* Bleeker	*Hydrocynus vittatus* Castelnau, 1861	
	Botswana	Okavango River and Delta	Okavango River and Delta	F	*Clarias gariepinus*	*Clarias gariepinus* (Burchell, 1822)	Van As & Van As (2015)
Argulus angusticeps Cunnington, 1913	Uvira	Lake Tanganyika	Lake Tanganyika	U	Not given	Not given	Cunnington (1913); Rushton-Mellor (1994b)
Argulus belones van Kampen, 1909[d]	KwaZulu Natal, South Africa	Not mentioned	Not given	M	*Sphyraena commersoni* (Barracuda)	*Sphyraena barracuda* (Edwards, 1771)	Barnard (1955)
Argulus brachypeltis Fryer, 1959	Zimbabwe	Lake Bangweulu	Lake Bangweulu	F	*Hydrocyon lineatus*	*Hydrocynus vittatus* Castelnau, 1861	Fryer (1959)
	Khartoum, Egypt	River Nile	Nile River	F	Free swimming	Free swimming	

TABLE V
(Continued)

Argulus species	Country/city/town	Water system as published	Water system confirmed[a]	E	Host species	Host species confirmed[b]	Reference
	Kenya	Lake Turkana	Lake Turkana	F	Free swimming	Free swimming	Rushton-Mellor (1994a)
Argulus capensis Barnard, 1955	Bredasdorp District, Cape Town, South Africa	Zoetendals Vlei	Soetendals Vlei	F	*Sandelia capensis* (Cape Kurper) (Cuvier & Valenciennes, 1831)	*Sandelia capensis* (Cuvier, 1829)	Barnard (1955)
Argulus cunningtoni Fryer, 1965	Not given	Lake Albert	Lake Albert	F	*Auchenoglanis occidentalis* (Cuvier & Valenciennes	*Auchenoglanis occidentalis* (Valenciennes, 1840)	Fryer (1965a)
	Not given	Lake Albert	Lake Albert	F	*Bagrus bayad* (Forskal)	*Bagrus bajad* (Forsskål, 1775)	
	Not given	Lake Albert	Lake Albert	F	*Clarias lazera* Cuvier & Valenciennes	*Clarias gariepinus* (Burchell, 1822)	
	Not given	Lake Albert	Lake Albert	F	*Synodontis schall* (Bloch-Schneider)	*Synodontis schall* (Bloch & Schneider, 1801)	
	Not given	Lake Albert	Lake Albert	F	*Lates niloticus* (L.)	*Lates niloticus* (Linnaeus, 1758)	
	Not given	Lake Albert	Lake Albert	F	*Distichodus niloticus* (L.)	*Distichodus nefasch* (Bonnaterre, 1788)	
	Not given	Lake Rudolf	Lake Turkana	F	*Lates* sp.	*Lates* sp.	Fryer (1968)
	Namibia	Zambezi River	Zambezi River	F	*Clarias gariepinus*	*Clarias gariepinus* (Burchell, 1822)	Van As & Van As (2015)

TABLE V
(Continued)

Argulus species	Country/city/town	Water system as published	Water system confirmed[a]	E	Host species	Host species confirmed[b]	Reference
Argulus dactylopteri Thorell, 1864	Not given	East Indian Ocean	East Indian Ocean	F	Serranochromis robustus	Serranochromis robustus (Günther, 1864)	Wilson (1902)
Argulus dageti Dollfus, 1960	Mopti, Mali	Niger River	Niger River	M	Dactylopterus volitanus Linnaeus	Dactylopterus volitans (Linnaeus, 1758)	Dollfus (1960)
				F	Heterobranchus bidorsalis	Heterobranchus bidorsalis Geoffroy Saint-Hilaire, 1809	
				F	Clarias anguillaris	Clarias anguillaris (Linnaeus, 1758)	
	Diafarabé, Mali	Niger River	Niger River	F	Tetrodon fahaka strigosus	Tetraodon lineatus Linnaeus, 1758	
				F	Tetrodon fahaka strigosus	Tetraodon lineatus Linnaeus, 1758	
Argulus dartevellei Brian, 1940	Point Padron	Unknown	Unknown	M	Polynemus quadrifilis	Polydactylus quadrifilis (Cuvier, 1829)	Brian (1940)
Argulus exiguus Cunnington, 1913	French Equatorial Africa	River Kouilou	Kouilou-Niari River	F	Promicrops distalis Roux & Calligan	Promicrops distalis Roux & Calligan	Fryer (1960a)
	Mpala	Lake Tanganyika	Lake Tanganyika	F	Simochromis diagramma	Simochromis diagramma (Günther, 1894)	Cunnington (1913); Rushton-Mellor (1994b)

TABLE V
(Continued)

Argulus species	Country/city/town	Water system as published	Water system confirmed[a]	E	Host species	Host species confirmed[b]	Reference
Argulus fryeri Rushton-Mellor, 1994	Mpala	Lake Tanganyika	Lake Tanganyika	F	Haplochilus tanganicanus	Lamprichthys tanganicanus (Boulenger, 1898)	Rushton-Mellor (1994a)
	Kenya	Lake Turkana	Lake Turkana	F	Host unknown	Host unknown	Otachi et al. (2015)
	Kenya	Near Kalokol, Longech spit (peninsula), eastern bank of the Ferguson's Gulf 03°33.218′N 035°54.742′E — Lake Turkana	Near Kalokol, Longech spit (peninsula), eastern bank of the Ferguson's Gulf 03°33.218′N 035°54.742′E — Lake Turkana	F	Tilapia zillii (Gervais, 1848)	Tilapia zillii (Gervais, 1848)	
Argulus giganteus Lucas, 1849[ef]	Algiers, Algeria	Algiers harbour	Algiers harbour	M	Skate or ray	Host unknown	Lucas (1849)
Argulus gracilis Rushton-Mellor, 1994		Lake Tanganyika	Lake Tanganyika	F	Auchenoglanis occidentalis var. tanganicanus	Auchenoglanis occidentalis (Valenciennes, 1840)	Rushton-Mellor (1994a)
Argulus incisus Cunnington, 1913	Rumonge	Lake Tanganyika	Lake Tanganyika	F	Auchenoglanis occidentalis var. tanganicanus	Auchenoglanis occidentalis (Valenciennes, 1840)	Cunnington (1913); Rushton-Mellor (1994b)

TABLE V
(Continued)

Argulus species	Country/city/town	Water system as published	Water system confirmed[a]	E	Host species	Host species confirmed[b]	Reference
Argulus izintwala Van As & Van As, 2001	KwaZulu Natal, South Africa	Lake St. Lucia	Lake St. Lucia	B	Hilsa kelee (Cuvier)	Hilsa kelee (Cuvier, 1829)	Van As & Van As (2001)
Argulus japonicus Thiele, 1900[g]	Alexandria, Egypt	1st Institute of Hydrobiology and fisheries	Mediterranean Sea	M	Royal carp	Royal carp	Fryer (1960a)
		Bloemhof Dam 27°40'S 26°0'E	Bloemhof Dam 27°40'S 26°0'E	F	Barbus holubi Steindachner	Labeobarbus aeneus (Burchell, 1822)	Kruger et al. (1983)
		Bloemhof Dam 27°40'S 26°0'E	Bloemhof Dam 27°40'S 26°0'E	F	Barbus kimberleyensis Gilchrist& Thompson	Labeobarbus kimberleyensis (Gilchrist & Thompson, 1913)	
		Bloemhof Dam 27°40'S 26°0'E	Bloemhof Dam 27°40'S 26°0'E	F	Clarias gariepinus (Burchell)	Clarias gariepinus (Burchell, 1822)	
		Bloemhof Dam 27°40'S 26°0'E	Bloemhof Dam 27°40'S 26°0'E	F	Cyprinus carpio L.	Cyprinus carpio Linnaeus, 1758	
		Bloemhof Dam 27°40'S 26°0'E	Bloemhof Dam 27°40'S 26°0'E	F	Labeo capensis (Smith)	Labeo capensis (Smith, 1841)	
		Bloemhof Dam 27°40'S 26°0'E	Bloemhof Dam 27°40'S 26°0'E	F	Labeo umbratus (Smith)	Labeo umbratus (Smith, 1841)	
		Lake Barberspan 26°35'S 25°35'E	Lake Barberspan 26°35'S 25°35'E	F	Barbus holubi Steindachner	Labeobarbus aeneus (Burchell, 1822)	

TABLE V
(Continued)

Argulus species	Country/city/town	Water system as published	Water system confirmed[a]	E	Host species	F	Host species confirmed[b]	Reference
		Lake Barberspan 26°35'S 25°35'E	Lake Barberspan 26°35'S 25°35'E	E	F	Clarias gariepinus (Burchell)	Clarias gariepinus (Burchell, 1822)	
		Lake Barberspan 26°35'S 25°35'E	Lake Barberspan 26°35'S 25°35'E		F	Labeo capensis (Smith)	Labeo capensis (Smith, 1841)	
		Lake Barberspan 26°35'S 25°35'E	Lake Barberspan 26°35'S 25°35'E		F	Labeo umbratus (Smith)	Labeo umbratus (Smith, 1841)	
		Lake Barberspan 26°35'S 25°35'E	Lake Barberspan 26°35'S 25°35'E		F	Tilapia sparrmanii (Smith)	Tilapia sparrmanii Smith, 1840	
	Ventersdorp, North-West Province, South Africa	Klerkskraal Dam	Klerkskraal Dam		F	Labeo umbratus (Smith, 1841)	Labeo umbratus (Smith, 1841)	Lutsch & Avenant-Oldewage (1995)
	Bronkhorstspruit, South Africa	Bronkhorstspruit Dam, Olifants River	Bronkhorstspruit Dam		F	Barbus marequensis	Labeobarbus marequensis (Smith, 1841)	Avenant-Oldewage (2001)
	Olifants River Lodge	Olifants River	Olifants River		F	Labeo umbratus	Labeo umbratus (Smith, 1841)	
					F	Clarias gariepinus	Clarias gariepinus (Burchell, 1822)	

TABLE V
(Continued)

Argulus species	Country/city/town	Water system as published	Water system confirmed[a]	E	Host species	Host species confirmed[b]	Reference
				F	Cyprinus carpio	Cyprinus carpio Linnaeus, 1758	
	Witbank, South Africa	Witbank Dam, Olifants River	Witbank Dam, Olifants River	F	Labeo umbratus	Labeo umbratus (Smith, 1841)	
				F	Clarias gariepinus	Clarias gariepinus (Burchell, 1822)	
	Potchefstroom, South Africa	Loskop Dam, Olifants River	Loskop Dam, Olifants River	F	Clarias gariepinus	Clarias gariepinus (Burchell, 1822)	
				F	Oreochromis mossambicus	Oreochromis mossambicus (Peters, 1852)	
	Deneysville, South Africa	Vaal Dam 26°52.249'S 28°10.249'E	Vaal Dam 26°52.249'S 28°10.249'E	F	Labeobarbus aeneus	Labeobarbus aeneus (Burchell, 1822)	Tam & Avenant-Oldewage (2006, 2009a)
				F	Labeobarbus kimberleyensis	Labeobarbus kimberleyensis (Gilchrist & Thompson, 1913)	
Argulus jollymani Fryer, 1956[h]		Nkata Bay, Lake Nyasa	Nkhata Bay, Lake Malawi	F	Haplochromis sp.	Haplochromis sp.	Fryer (1956)
		Ruarwe, Lake Nyasa	Lake Malawi	F	Haplochromis fenestratus Trewavas	Protomelas fenestratus (Trewavas, 1935)	
Argulus kosus Avenant-Oldewage, 1994	KwaZulu Natal, South Africa	Kosi Bay	Kosi Bay estuary	B	Sarpa salpa Linn. (strepie)	Sarpa salpa (Linnaeus, 1758)	Avenant-Oldewage (1994a)

TABLE V
(Continued)

Argulus species	Country/city/town	Water system as published	Water system confirmed[a]	E	Host species	Host species confirmed[b]	Reference
	KwaZulu Natal, South Africa	Lake St. Lucia 28°10'S 32°30'E	Lake St. Lucia 28°10'S 32°30'E	B	Ladyfish *Elops machnata* (Forsskål)	*Elops machnata* (Forsskål, 1775)	Van As et al. (1999)
				B	St Lucia mullet *Liza luciae* (Penrith & Penrith)	*Liza luciae* (Penrith & Penrith, 1967)	
				B	flat- head mullet *Mugil cephalus* Linnaeus	*Mugil cephalus* Linnaeus, 1758	
				B	Mozambique tilapia *Oreochromis mossambicus* (Peters)	*Oreochromis mossambicus* (Peters, 1852)	
				B	snapper kob *Otolithes ruber* (Schneider)	*Otolithes ruber* (Bloch & Schneider, 1801)	
				B	spotted grunter *Pomadasys commersonni* (Lacépède	*Pomadasys commersonnii* (Lacépède, 1801)	
				B	cock grunter *Pomadasys multimaculatum* (Playfair)	*Pomadasys multimaculatus* (Playfair, 1867)	
				B	Cape stump-nose *Rhabdosargus holubi* (Steindachner)	*Rhabdosargus holubi* (Steindachner, 1881)	

TABLE V
(Continued)

Argulus species	Country/city/town	Water system as published	Water system confirmed[a]	E	Host species	E	Host species confirmed[b]	Reference
Argulus melita Van Beneden, 1891	Senegal	Bay of Dakar	Atlantic Ocean	M	Shark	M	Shark	Van Beneden (1891)
	Senegal	Bay of Dakar	Atlantic Ocean	M	Shark	M	Shark	Wilson (1902)
Argulus monodi Fryer, 1959	Zimbabwe	Lake Bangweulu	Lake Bangweulu	F	*Hydrocyon lineatus*	F	*Hydrocynus vittatus* Castelnau, 1861	Fryer (1959)
	Kenya	Near Kalokol, Longech spit (peninsula), eastern bank of the Ferguson's Gulf 03°33.218′N 035°54.742′E — Lake Turkana	Near Kalokol, Longech spit (peninsula), eastern bank of the Ferguson's Gulf 03°33.218′N 035°54.742′E — Lake Turkana	F	*Tilapia zillii* (Gervais, 1848)	F	*Tilapia zillii* (Gervais, 1848)	Otachi et al. (2015)
Argulus multipocula Barnard, 1955	KwaZulu Natal, South Africa	Richards Bay	Richards Bay	B	Not given	B	Not given	Barnard (1955)
	West Coast, South Africa	Berg River Estuary	Berg River Estuary	B	*Liza richardsonii* (Smith) Southern Mullet	B	*Liza richardsonii* (Smith, 1846)	Smit et al. (2005)
Argulus otolithi Brian, 1927[i]	Cameroon	Kribi	Gulf of Guinea	M	*Pseudotolithus typus*	M	*Pseudotolithus typus* Bleeker, 1863	Brian (1927)
Argulus personatus Cunnington, 1913	Ndanvie	Lake Tanganyika	Lake Tanganyika	F	*Bathybates ferox*	F	*Bathybates ferox* Boulenger, 1898	Cunnington (1913); Rushton-Mellor (1994b)

TABLE V
(Continued)

Argulus species	Country/city/town	Water system as published	Water system confirmed[a]	E	Host species	Host species confirmed[b]	Reference
	Tanzania	Lake Tanganyika	Lake Tanganyika	F	*Bathybates fasciatus* Boulenger	*Bathybates fasciatus* Boulenger (1901)	Fryer (1965a)
	Mpulunga, northern Zambia	Lake Tanganyika	Lake Tanganyika	F	*Bathybates ferox* Boulenger, 1898	*Bathybates ferox* Boulenger, 1898	Tam et al. (2005)
Argulus reticulatus Wilson, 1920	Malela, Congo	Congo River	Congo River	F	*Hydrocyon goliath* Boulenger	*Hydrocynus goliath* Boulenger, 1898	Wilson (1920b)
Argulus rhipidiophorus Monod, 1931	Kawa, Republic of Congo	Lake Albert	Lake Albert	F	*Hydrocyon* sp.	*Hydrocyon* sp.	Monod (1931)
	Not given	Lake Rudolf	Lake Turkana	F	*Clarias lazera* Cuvier & Valenciennes	*Clarias gariepinus* (Burchell, 1822)	Fryer (1960a)
	Not given	Lake No on the upper White Nile	Lake No, White Nile	F	Not given	Not given	
	Kenya	Lake Naivasha	Lake Naivasha	F	Not given	Not given	
	Bitshumbi	Lake Edward	Lake Edward	F	*Tilapia nilotica regani* Poll.	*Oreochromis niloticus* (Linnaeus, 1758)	Fryer (1963)
	Bitshumbi	Lake Edward	Lake Edward	F	*Tilapia leucosticta* Trewavas	*Oreochromis leucostictus* (Trewavas, 1933)	
	Not given	Lake Edward	Lake Edward	F	*Haplochromis pappenheimi* (Boulenger)	*Haplochromis pappenheimi* (Boulenger, 1914)	Fryer (1965a)

TABLE V
(Continued)

Argulus species	Country/city/town	Water system as published	Water system confirmed[a]	E	Host species	Host species confirmed[b]	Reference
	Not given	Lake Kivu	Lake Kivu	F	Tilapia nilotica regani Poll.	Oreochromis niloticus (Linnaeus, 1758)	Fryer (1965a)
	Not given	Lake Kivu	Lake Kivu	F	Tilapia nilotica regani Poll.	Oreochromis niloticus (Linnaeus, 1758)	Fryer (1965b)
	Not given	Lake Edward and the Kazinga Channel	Lake Edward and the Kazinga Channel	F	Haplochromis sp.	Haplochromis sp.	
	Not given	Lake Albert	Lake Albert	F	Hydrocyon lineatus Bleeker	Hydrocynus vittatus Castelnau, 1861	
	Not given	Lake Albert	Lake Albert	F	Hydrocyon forskali Cuvier	Hydrocynus forskahlii (Cuvier, 1819)	
	Not given	Lake Albert	Lake Albert	F	Alestes baramose (Joannis)	Alestes baremoze (Joannis, 1835)	
	Not given	Lake Albert	Lake Albert	F	Lates niloticus (L.)	Lates niloticus (Linnaeus, 1758)	
	Not given	Lake Albert	Lake Albert	F	Synodontis schall (Bloch-Schneider)	Synodontis schall (Bloch & Schneider, 1801)	
	Not given	Lake Albert	Lake Albert	F	Bagrus bayad (Forskal)	Bagrus bajad (Forsskål, 1775)	
	Not given	Lake Albert	Lake Albert	F	Clarias lazera Cuvier & Valenciennes	Clarias gariepinus (Burchell, 1822)	

TABLE V
(Continued)

Argulus species	Country/city/town	Water system as published	Water system confirmed[a]	E	Host species	Host species confirmed[b]	Reference
	Khartoum, Egypt	River Nile	Nile River	F	Free swimming	Free swimming	Fryer (1968)
	Ethiopia	Lake Zwai (Zwei)	Lake Zway	F	Tilapia sp. Barbus sp.	Tilapia sp. Barbus sp.	Brian (1940)
	Ethiopia	Lake Oitu Lake Langano)	Lake Langano	F	Tilapia sp. Barbus sp.	Tilapia sp. Barbus sp.	
	Ethiopia	Lake Awassa (Awusa)	Lake Awasa	F	Clarias sp.	Clarias sp.	
	Ethiopia	Lake Abaya (Lake Margherita)	Lake Abaya	F / F	Clarias sp. Labeo sp.	Clarias sp. Labeo sp.	
Argulus rijckmansii Brian, 1940	Not given	Lake Rudolf	Lake Turkana	F	Lates sp.	Lates sp.	
	Ango-Ango, Democratic Republic of Congo	Congo River	Congo River	F	Not given	Not given	
Argulus rubescens Cunnington, 1913	Moliro	Lake Tanganyika	Lake Tanganyika	F	Chrysichthys brachynema	Chrysichthys brachynema Boulenger (1900)	Cunnington (1913); Rushton-Mellor (1994b)
	Kibwesi	Lake Tanganyika	Lake Tanganyika	F	Chrysichthys brachynema	Chrysichthys brachynema Boulenger (1900)	

TABLE V
(Continued)

Argulus species	Country/city/town	Water system as published	Water system confirmed[a]	E	Host species	Host species confirmed[b]	Reference
Argulus rubropunctatus Cunnington, 1913	Mrumbi	Lake Tanganyika	Lake Tanganyika	F	Chrysichthys brachynema	Chrysichthys brachynema Boulenger (1900)	Cunnington (1913); Rushton-Mellor (1994b)
	Kibwesi	Lake Tanganyika	Lake Tanganyika	F	Lates microlepis	Lates microlepis Boulenger, 1898	
	Masawa	Lake Tanganyika	Lake Tanganyika	F	Lates microlepis	Lates microlepis Boulenger, 1898	
	Ndanvie	Lake Tanganyika	Lake Tanganyika	F	Lates microlepis	Lates microlepis Boulenger, 1898	
	Not given	Lake Tanganyika	Lake Tanganyika	F	Lates microlepis Boulenger	Lates microlepis Boulenger, 1898	Fryer (1965a)
Argulus schoutedeni Monod, 1928	Bukama	Katanga River	Luvua River	F	Not given	Not given	Monod (1928)
	Ikela	River Chuapa (Tshuapa), Congo System	Tshuapa River	F	Distochodus fasciolatus Boulenger	Distichodus fasciolatus Boulenger, 1898	Fryer (1964)
	Not given	Lake Tanganyika	Lake Tanganyika	F	Citharinus gibbosus Boulenger	Citharinus gibbosus Boulenger, 1899	Fryer (1965a)
Argulus smalei Avenant-Oldewage & Oldewage, 1995[j]	Port Elizabeth, South Africa	Algoa Bay	Algoa Bay	M	Aluterus monoceros (Linn., 1758), unicorn leatherjacket	Aluterus monoceros (Linnaeus, 1758)	Avenant-Oldewage & Oldewage (1995)

TABLE V
(Continued)

Argulus species	Country/city/town	Water system as published	Water system confirmed[a]	E	Host species	F	Host species confirmed[b]	Reference
Argulus striatus Cunnington, 1913	Mbete	Lake Tanganyika	Lake Tanganyika	F	*Dinopterus cunningtoni*		*Dinotopterus cunningtoni* Boulenger (1906)	Cunnington (1913); Rushton-Mellor (1994b)
	Sumbu	Lake Tanganyika	Lake Tanganyika	F	*Clarias robecchii*		*Clarias gariepinus* (Burchell, 1822)	
	Vua	Lake Tanganyika	Lake Tanganyika	F	*Clarias lazera* Muomi		*Clarias gariepinus* (Burchell, 1822)	
	Kasawa	Lake Tanganyika	Lake Tanganyika	F	*Dinopterus cunningtoni*		*Dinotopterus cunningtoni* Boulenger (1906)	
	Kibwesi	Lake Tanganyika	Lake Tanganyika	F	*Chrysichthys brachynema*		*Chrysichthys brachynema* Boulenger (1900)	
	Rumonge	Lake Tanganyika	Lake Tanganyika	F	*Auchenoglanis occidentalis var. tanganicanus*		*Auchenoglanis occidentalis* (Valenciennes, 1840)	
	unknown	Lake Tanganyika	Lake Tanganyika	F	*Auchenoglanis occidentalis* (Cuvier & Valenciennes)		*Auchenoglanis occidentalis* (Valenciennes, 1840)	Fryer (1965a)
	unknown	Lake Tanganyika	Lake Tanganyika	F	*Tilapia tanganicae* (Günther)		*Oreochromis tanganicae* (Günther, 1894)	

TABLE V
(Continued)

Argulus species	Country/city/town	Water system as published	Water system confirmed[a]	E	Host species	Host species confirmed[b]	Reference
	unknown	Lake Tanganyika	Lake Tanganyika	F	*Heterobranchus* sp.	*Heterobranchus* sp.	
	unknown	Lake Tanganyika	Lake Tanganyika	F	Host unknown	Host unknown	
	unknown	Lake Tanganyika	Lake Tanganyika	F	*Dinopterus cunningtoni* Boulenger	*Dinotopterus cunningtoni* Boulenger (1906)	
Argulus trachynoti Brian, 1927	Souelaba, Cameroon	Douala Bay	Douala Bay	M	*Trachynotus falcatus* Linne	*Trachinotus blochii* (Lacépède, 1801)	Brian (1927)
Argulus vittatus (Rafinesque-Schmaltz, 1814)[k]	Not given	Not given	Not given	M	*Sparus crythrinus*	*Pagellus erythrinus* (Linnaeus, 1758)	Rafinesque-Schmaltz (1814)
	East coast of Algeria	Gulf of Béjaïa	Gulf of Béjaïa	M	*Boops boops*	*Boops boops* (Linnaeus, 1758)	Ramdane & Trilles (2012)
	East coast of Algeria	Gulf of Béjaïa	Gulf of Béjaïa	M	*Pagrus pagrus*	*Pagrus pagrus* (Linnaeus, 1758)	
	East coast of Algeria	Gulf of Béjaïa	Gulf of Béjaïa	M	*Boops boops* (L., 1758)	*Boops boops* (Linnaeus, 1758)	Ider et al. (2014)
	East coast of Algeria	Gulf of Béjaïa	Gulf of Béjaïa	M	*Sparus aurata* (L., 1758)	*Sparus aurata* Linnaeus, 1758	
	East coast of Algeria	Gulf of Béjaïa	Gulf of Béjaïa	M	*Pagellus erythrinus* (L., 1758)	*Pagellus erythrinus* (Linnaeus, 1758)	
Argulus wilsoni Brian, 1940	Boma, Democratic Republic of Congo	Kalamu River	Kalamu River	F	*Hydrocyon goliath*	*Hydrocynus goliath* Boulenger, 1898	Brian (1940)

TABLE V
(Continued)

Argulus species	Country/city/town	Water system as published	Water system confirmed[a]	E	Host species	Host species confirmed[b]	Reference
Argulus zei Brian, 1924[l]	Republic of Mauratania	Not given	Not given	M	Zeus faber	Zeus faber Linnaeus, 1758	Brian (1924)

E, environment in which it is found: B, brackish water; F, freshwater; M, marine; U, unknown.

[a]) Water system confirmed.

[b]) Host species confirmed using fishbase.org.

[c]) Male of A. ambloplites Wilson, 1920 is designated A. confusus Rushton-Mellor (1994a).

[d]) First described from Asia.

[e]) A. giganteus Lucas, 1849 synonym to A. purpureus Risso according to Wilson (1902) and Rushton-Mellor (1994b).

[f]) A. giganteus Lucas, 1849 synonym of A. vittatus according to Boxshall & Walter (2009) and Ramdane & Trilles (2012).

[g]) Cosmopolitan distribution.

[h]) A. jollymani Fryer, 1956 changes to A. ambloplites jollymani (Fryer, 1960a).

[i]) A. otolithi Brian, 1927 synonym of A. arcassonensis Cuénot (1912) according to Rushton-Mellor (1994b).

[j]) A. smalei Avenant-Oldewage & Oldewage, 1995 synonym of A. kosus according to Van As et al. (1999).

[k]) A. vittatus (Rafinesque-Schmaltz, 1814) is a senior synonym of A. purpureus (Risso, 1827) according to Holthuis (1954) and Poly (1998b).

[l]) A. zei Brian, 1924 synonym of A. arcassonensis Cuénot, 1912 according to Monod (1928).

TABLE VI

Marine and freshwater *Argulus* spp. from Australia and Oceania

Argulus species	Country/city/town	Water system as published	Water system confirmed[a]	E	Host species	Host species confirmed[b]	Reference
Argulus australiensis Byrnes, 1985	Kurumba, Queensland	Unknown	Arafura sea, Indian Ocean	M	*Acanthopagrus berda*	*Acanthopagrus berda* (Forsskål, 1775)	Byrnes (1985)
Argulus diversicolor Byrnes, 1985	Carnarvon, Western Australia	Point Samson	Point Samson, Indian Ocean	M	*Acanthopagrus latus*	*Acanthopagrus latus* (Houttuyn, 1782)	Byrnes (1985)
Argulus japonicus Thiele, 1900[c]	Sydney, New South Wales	Aquarium	Aquarium	F	*Carassius auratus*	*Carassius auratus* (Linnaeus, 1758)	Heegard (1962)
	New Zealand	Aquarium	Aquarium	F	Goldfish imported from Southeast Asia	*Carassius auratus* (Linnaeus, 1758) imported from Southeast Asia	Pilgrim (1967)
Argulus macropterus Heegard, 1962	Mandurah, Western Australia	Murray River	Murray River	F	*Mugil* sp.	*Mugil* sp.	Heegard (1962)
Argulus papuensis Rushton-Mellor, 1991	Port Moresby, Papua New Guinea	Sepik River	Sepik River	F	*Bunaka herwerdeni* (Weber)	*Buraka gyrinoides* (Bleeker, 1853)	Rushton-Mellor (1991)

E, environment in which it is found: F, freshwater; M, marine.

[a] Water system confirmed.

[b] Host species confirmed using fishbase.org.

[c] Cosmopolitan distribution.

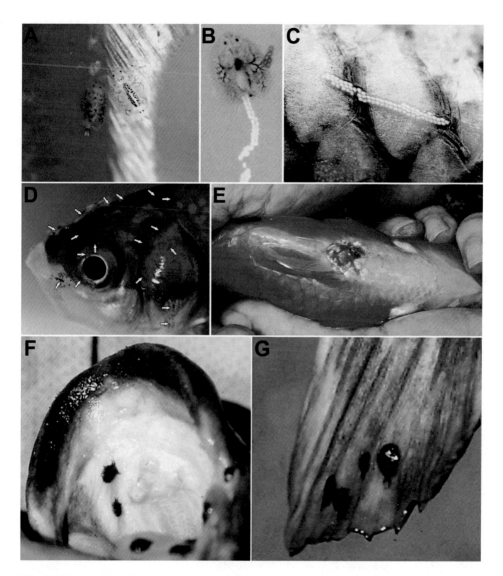

Fig. 5. Micrographs of Branchiura and their hosts. A, a copulating pair (left) and post-copulatory pair (right) of *Argulus japonicus* Thiele, 1900 on *Carassius auratus* (Linnaeus, 1758); B, a female *Argulus japonicus* Thiele, 1900 depositing eggs on the tank wall; C, *Argulus japonicus* Thiele, 1900 eggs deposited on the body of a dead *Labeobarbus aeneus* (Burchell, 1822); D, *Argulus japonicus* Thiele, 1900 distributed on *Carassius auratus* (Linnaeus, 1758), arrows indicate each argulid; E, a female *Argulus japonicus* Thiele, 1900 and the lesion caused on *Carassius auratus* (Linnaeus, 1758); F, *Dolops ranarum* (Stuhlmann, 1891) in the mouth of *Oreochromis mossambicus* (Peters, 1852); G, *Dolops ranarum* (Stuhlmann, 1891) on the fin of *Clarias gariepinus* (Burchell, 1822). [Figs. F and G courtesy Johan Theron.]

TABLE VII

Summary of selected records of research conducted on the anatomy and physiology that have been conducted on *Argulus* species

	Species	References
Digestive system	*Argulus* spp.	Claus (1875); Wilson (1902)
	A. coregoni	Shimura (1983b); Shimura & Inoue (1984)
	A. foliaceus	Jurine (1806); Leydig (1850); Grobben (1908); Debaisieux (1953); Madsen (1964); Czeczuga (1971)
	A. japonicus	Shimura (1983b); Swanepoel & Avenant-Oldewage (1992); Gresty et al. (1993); Tam & Avenant-Oldewage (2006, 2009a); Walker et al. (2011b)
	A. siamensis	Saha et al. (2011)
	A. viridis (= *A. foliaceus*)	Martin (1932)
Nervous system	*Argulus* spp.	Wilson (1902)
	A. americanus	Wilson (1904)
	A. foliaceus	Leydig (1850); Claus (1875); Zaćwilichowska (1948); Debaisieux (1953); Madsen (1964); Van den Bosch de Aguilar (1972)
	A. viridis (= *A. foliaceus*)	Martin (1932)
Sensory system	*Argulus* spp.	Walker et al. (2004)
	A. coregoni	Meyer-Rochow et al. (2001); Bandilla et al. (2007)
	A. foliaceus	Leydig (1850); Debaisieux (1953); Madsen (1964); Hallberg (1982); Mikheev et al. (1998, 2000); Meyer-Rochow et al. (2001)
	A. japonicus	Yoshizawa & Nogami (2008)
	A. viridis (= *A. foliaceus*)	Martin (1932)
Respiratory system	*Argulus* spp.	Thorell (1864); Wilson (1902); Fryer (1968)
	A. foliaceus	Jurine (1806); Leydig (1850); Claus (1875); Jírovec & Wenig (1934); Fox (1957); Haase (1974); Haase (1975a, b)
Circulatory system	*Argulus* spp.	Wilson (1902)
	A. americanus	Wilson (1904)
	A. foliaceus	Debaisieux (1953)
	A. japonicus	Leydig (1850); Kondo et al. (2003)
Reproductive system male	*Argulus* spp.	Claus (1875); Wilson (1902)
	A. americanus	Wilson (1904)
	A. bengalensis	Banerjee et al. (2014a)
	A. coregoni	Pasternak et al. (2004b)
	A. foliaceus	Leydig (1850); Debaisieux (1953); Wingstrand (1972)
	A. japonicus	Avenant-Oldewage & Swanepoel (1993); Avenant-Oldewage & Everts (2010)
Reproductive system female	*Argulus* spp.	Claus (1875); Wilson (1902)
	A. bengalensis	Banerjee & Saha (2016)
	A. americanus	Wilson (1904)
	A. coregoni	Pasternak et al. (2004b)

TABLE VII
(Continued)

	Species	References
	A. foliaceus	Leydig (1850); Debaisieux (1953); Kubrakiewicz & Klimowicz (1994)
	A. japonicus	Ikuta & Makioka (1992, 1994, 1997); Ikuta et al. (1997)
	A. viridis (= A. foliaceus)	Martin (1932)
Life cycle	Argulus spp.	Claus (1875); Kellicott (1880); Wilson (1902); Stammer (1959); Bauer (1962); Fryer (1964); Sundara Bai et al. (1988); Thatcher (1991, 2006); Walker et al. (2004); Schram et al. (2005); Møller et al. (2008)
	A. americanus	Wilson (1904)
	A. bengalensis	Guha et al. (2012, 2013); Banerjee et al. (2015)
	A. coregoni	Shimura & Egusa (1980); Shimura (1981, 1983a); Hakalahti & Valtonen (2003); Hakalahti et al. (2004a, c, 2005); Pasternak et al. (2004a); Kaji et al. (2011)
	A. foliaceus	Leydig (1850); Schlüter (1979); Rushton-Mellor & Boxshall (1994); Pasternak et al. (2000); Harrison et al. (2006, 2007); Møller et al. (2007); Walker et al. (2008a); Taylor et al. (2009a, b); Koyun (2011)
	A. funduli	Wilson (1907)
	A. japonicus	Tokioka (1936b); Shafir & Van As (1986); Shafir & Oldewage (1992); Gresty et al. (1993); Lutsch & Avenant-Oldewage (1995)
	A. maculosus	Wilson (1907)
	A. megalops	Davis (1965)
	A. puthenveliensis	Thomas (1961)
	A. varians (= A. megalops)	Bouchet (1985)

Digestive system.— The digestive system of *Argulus* species has been repeatedly investigated with researchers providing details on the anatomy and histology (Jurine, 1806; Leydig, 1850; Madsen, 1964; Tam & Avenant-Oldewage, 2006, 2009a). Jurine (1806) described the digestive system simply as having an oesophagus, a stomach with lateral branches, a cecum and a rectum. Leydig (1850) later added that there are glands attached to the preoral spine. Much later, Madsen (1964) summarised the digestive system as consisting of the oesophagus, the crop, the enteral diverticula, the midgut, and the rectum. Recently, Tam & Avenant-Oldewage (2006) provided a histological description of the digestive system of larval *Argulus japonicus* Thiele, 1900. They found that the larvae feed mainly on mucous and epithelial cells of their fish host. Later, Tam & Avenant-Oldewage (2009a) provided a description of the ultrastructure of the adult digestive system. Their paper provided the basis for Saha et al. (2011) and Walker et al. (2011b) to prove that the digestive system of at least adult argulids functions to digest the blood of its fish host. This had long since been speculated with researchers such as Shimura et al. (1983b) and Shimura & Inoue (1984) already having posited this thirty years before.

Nervous system.— As with the digestive system, the nervous system of *Argulus* species seems well documented. Martin (1932) was the first to describe that the supra-oesophageal ganglion mass can be differentiated into three brain masses; the protocerebrum leading to the optic ganglia (eyesight), deuterocerebrum with ganglia leading to the antennules, the naupliar eye and the preoral spine, and tritocerebrum with ganglia leading to the antennae, the gland of the preoral spine and the ventral side of the carapace. Zaćwilichowski (1935) found that the antennulae, antennae and maxillulae have ganglia that originate from the deuterocerebrum. Thereafter, in a later paper, Zaćwilichowska (1948) contradicted Martin (1932) only in the number of cerebral and ventral ganglia present. She also provided a nerve map for *A. foliaceus* which extended to the abdomen of the species. Later, Madsen (1964) confirmed that the central nervous system consists of one pair of cerebral ganglia, one pair of suboesophageal ganglia and five pairs of ventral ganglia. Van den Bosch de Aguilar (1972) was the last to contribute and described the neurosecretory cells of the nervous system as consisting of large amounts of APS positive granules within the cytoplasm, which reacted to staining for APS.

Sensory system.— Most of the recent sensory studies on argulids have been targeted towards the role of sight in finding hosts and mates. Mikheev et al. (1998) found that *A. foliaceus* is attracted to flashes or sources of light and as such would attach to bright silver fish faster in a dark environment before a darker fish host. However, further research by Mikheev et al. (2000) showed that their vision is better in a lit surrounding allowing for an ambush strategy to attach to a host, while darker tanks prompted the parasites to swim and search for a host. The ultrastructural study by Meyer-Rochow et al. (2001) found that *A. foliaceus* has up to 60 ommatidia containing crystalline cones, while *A. coregoni* contains 90 ommatidia at most. These authors were also able to confirm the behavioural findings of Mikheev et al. (1998) and Mikheev et al. (2000). Later, Yoshizawa & Nogami (2008) confirmed that *Argulus* species have a unique circadian rhythm in comparison to other crustaceans, and that they respond to different hues of light in the order of blue light, yellow light, green light, and red light.

In a different approach, Bandilla et al. (2007) tested whether a hierarchy exists in the stimuli to which argulids respond. They tested whether chemical (olfactory) stimuli would rank higher in mate and host location to visual (optic) stimuli. The results interestingly showed that light and fish odour were of greater importance than mate recognition in both males and females, suggesting that the parasites' first priority is host location followed by mate location (Bandilla et al., 2007).

Respiratory system.— Very little work has been conducted on the respiratory system and how it functions. Thorell (1864) was the first to suggest that the fleshy abdomen was involved in respiration, a hypothesis which Fryer (1968) later agreed with, adding that it must work in combination with the respiratory areas on the carapace lobes. Meanwhile, Fox (1957) was unable to detect oxyhaemoglobin in the blood of *Argulus foliaceus* L., but reported the detection in blood of *D. ranarum*. Interestingly, Haase (1974) identified the respiratory areas on the carapace lobes as being made up of ion-transporting epithelia. Thereafter, Haase (1975a, b) proved that sodium and chloride ions are located within the cytoplasm and microridges of the cells of the respiratory areas, and that ATPase activity occurs in order to transport ions across the membrane. A study on oxygen transport within argulids is still lacking.

Circulatory system.— Jurine (1806) was the first to describe the heart of *Argulus* species as cylindrical and consisting of a single ventricle within a tubercle. This was followed later by Leydig (1850), Claus (1875) and Wilson (1902, 1904). Wilson (1902) described the heart as being triangular in outline, with the apex passing into the cylindrical aorta that leads to the brain and opens into the coelom. No further work has been conducted on the anatomy of the heart.

Recently, Kondo et al. (2003) conducted a morphological and cytochemical study on the hemocyte of *A. japonicus*. They found that the hemocyte is a granular ellipsoid cell containing glycogen and exhibits phenoloxidase activity; implying that the hemocyte has similar phagocytic and phenoloxidase properties to other crustaceans.

Reproductive system male.— Avenant-Oldewage & Everts (2010) provided a thorough literature review of the studies conducted on the male reproductive system in argulids in their study on the production of spermatophores in *A. japonicus*. Theirs was the first paper to describe the occurrence of spermatophores in the genus *Argulus*. Recently Banerjee et al. (2014a) described spermatogenesis and spermiogenesis in *Argulus bengalensis* Ramakrishna, 1951 using fluorescent light microscopy, scanning and transmission electron microscopy. This publication added to the information provided by Wingstrand (1972) for sperm morphology. Their paper also contradicted that of Avenant-Oldewage & Everts (2010) in that they erroneously reported that in *A. bengalensis* spermatophores are formed in the seminal vesicle of the male and attach to the oviduct wall of the female (Banerjee et al., 2014a), but they did not explain two key points; first, how the spermatophores are transferred to the female from the male; and second, how the female then inseminates the eggs when the sperm is housed in a spermatophore. Without further study, this paper (Banerjee et al., 2014a) may be cited for its incorrect interpretation of the male and female reproductive system.

Reproductive system female.— The female reproductive system in its simplest description, consists of an ovary with paired oviducts in the thorax, and seminal receptacles or spermathecae in the anterior half of the abdomen (Leydig, 1850; Claus, 1875; Wilson, 1902, 1904; Martin, 1932; Debaisieux, 1953). The last review of the reproductive system of the female was conducted by Ikuta & Makioka (1997) in which they describe the female system as a sac-like ovary with the walls of the ovary folded ventro-laterally. They also described how the eggs develop in two stages, the first on the outside of the ovarian wall and the second within the functional oviduct (Ikuta & Makioka, 1997). In a contradictory paper, Banerjee et al. (2014a) reported that the spermatheca is a misnomer as they did not find evidence of sperm stored within these structures in A. bengalensis; this despite all the previous literature describing the existence of this structure in other females of the genus Argulus, and micrographs in their own publication of the sperm stored in the spermathecae (see fig. 8 of Banerjee et al., 2014a).

Life cycle.— The life cycle of Argulus species has been extensively studied by various authors (see table VII). The number of larval stages that occur differs between species as Tokioka (1936b) only recorded 7 stages in A. japonicus development while Stammer (1959) and Shimura (1981) recorded 9 larval stages for A. foliaceus L., A. japonicus and A. coregoni Thorell, 1865 respectively. Similarly, Rushton-Mellor & Boxshall (1994) found 11 larval stages for A. foliaceus. The most prominent feature that has been identified is the maxillulae of first stage larvae which are modified to an elongate appendage that terminates in two hooks, one of these is serrated (Fryer, 1964; Møller et al., 2007). As development continues, a sucker forms in the second segment, and the distal segments reduce in size until only the sucker remains. Kaji et al. (2011) recently used confocal scanning electron microscopy to show the musculoskeletal ontogeny of the maxillules, proving definitively that the sucker forms mostly from the second segment.

Parasite-host interactions in Argulus species

Parasite-host interactions describe the mechanisms used by the species to locate a new host (linked to sensory system), effect on the host (behavioural, physical or secondary; linked to the digestive system), Argulus species as intermediate hosts for other parasites, and the control of this parasite (chemical, physical or biological). As with the section above, table VIII lists some of the research papers on these topics available to the authors.

Host location.— Host location has already been discussed in its link to the sensory system. It involves the use of sight and chemotaxis.

TABLE VIII

Summary of selected records of the research on parasite-host interactions conducted on *Argulus* species

	Species	References
Host location	*Argulus* spp.	Walker et al. (2004)
	A. coregoni	Mikheev et al. (2004, 2007, 2015); Bandilla et al. (2005, 2007, 2008)
	A. foliaceus	Mikheev et al. (1998, 2000, 2007, 2015); Walker et al. (2008b)
	A. japonicus	LaMarre & Cochran (1992); Yoshizawa & Nogami (2008)
Effect on host	*Argulus* spp.	Bower-Shore (1940); Bauer (1962); Hoffman (1977); Fryer (1978); Boomker (1981); Thatcher (1991, 2006); Moravec et al. (1999); Walker et al. (2004); Lester & Hayward (2006); Taylor et al. (2006); Akter et al. (2007); Webb (2008); Banerjee & Bandyopadhyay (2010); Uma et al. (2012); Kumar et al. (2016)
	A. africanus	Okaeme et al. (1988)
	A. biramosus	Allum & Hugghins (1959)
	A. canadensis	Dugatkin et al. (1994)
	A. coregoni	Tikhomirova (1970); Shimura et al. (1983a, b); Shimura & Inoue (1984); Bandilla et al. (2005, 2006); Mousavi et al. (2011); Mikheev et al. (2015)
	A. foliaceus	Tikhomirova (1970); Czeczuga (1971); Pfeil-Putzien (1977); Moravec (1978); Pfeil-Putzien & Baath (1978); Schlüter (1978); Tikhomirova (1980); Ahne (1985); Menezes et al. (1990); Molnár & Moravec (1997); Molnár & Székely (1998); Ruane et al. (1999); Yildiz & Kumantas (2002); Taylor et al. (2009a); Mikheev & Pasternak (2010); Mousavi et al. (2011); Pekmezci et al. (2011); Møller (2012); Mikheev et al. (2015)
	A. indicus	Singhal et al. (1990)
	A. japonicus	Swanepoel & Avenant-Oldewage (1992); Gresty et al. (1993); Watson & Avenant-Oldewage (1996); van der Salm et al. (2000); Haond et al. (2003); Tam & Avenant-Oldewage (2006, 2009a); Forlenza et al. (2008); Mousavi et al. (2011); Sahoo et al. (2012)
	A. siamensis	Nandp & Das (1991); Sahoo et al. (2012)
	A. trilineatus	Guberlet (1928)
Control	*Argulus* spp.	Hindle (1948); Stammer (1959); Merla (1961); Bauer (1962); Hines (1972); Hoffman (1977); Puffer & Beal (1981); Thatcher (1991, 2006); Benz et al. (2001); Wolfe et al. (2001); Walker et al. (2004, 2011a); Chowdhury et al. (2006); Hanson et al. (2011); Kumar et al. (2013, 2016); Parvez et al. (2013); Merk (2016)
	A. bengalensis	Banerjee et al. (2014b)
	A. coregoni	Inoue et al. (1980); Mikheev et al. (2001, 2015); Hakalahti et al. (2004b, c); Fenton et al. (2006)
	A. foliaceus	Chen (1933); Ruane et al. (1995); Gault et al. (2002); Harrison et al. (2006, 2007); Møller (2012); Mikheev et al. (2015)
	A. indicus	Singhal et al. (1986)
	A. japonicus	Vondrka (1972); Han et al. (1998)
	A. siamensis	Saurabh et al. (2010); Hemaprasanth et al. (2012); Sahoo et al. (2013)
	A. trilineatus	Guberlet (1928)

Effect on host.— The effect which this genus has on its host has been extensively investigated (see table VIII).

Bower-Shore (1940) suggested that *A. foliaceus* causes destruction to the colour pigments of stickleback fish, and that infection by *A. foliaceus* damages the mucosal layer making the affected fish susceptible to secondary infection with fungi. Shortly thereafter, Allum & Hugghins (1959) described the severe epizootics of an outbreak of *Argulus biramosus* Bere, 1931 in Lake Brant and a less severe outbreak in Lake Poinsett. In their description, Allum & Hugghins (1959) list inflammation and cytolytic damage to the scale pockets of the dermal scales of the fish as well as secondary infections by *Saproglenia* as signs of a severe infection. A chromatographic study of the caretenoids present in the fish louse, *A. foliaceus*, and their host, *Gasterosteus aculeatus* L., showed that the caretenoids present in both host and parasite differed in that not all caretenoids found in the tissue of the host were assimilated by *A. foliaceus*, and that some found in *A. foliaceus* were not present in the host tissue (Czeczuga, 1971). This proved that *A. foliaceus* only absorbs certain metabolites from the host tissue. Shimura et al. (1983b) studied the haematological changes caused by *A. coregoni* on trout (*Oncorhynchus masou* Brevoort, 1856) and noted significant changes in infected fish. Thereafter, Shimura & Inoue (1984) reported that toxic effects from the mouth extract of *A. coregoni* elicits a haemorrhagic response that aids in the sucking of blood from the host and may cause secondary anaemia but that the extract is neither haemolytic nor cytotoxic. Swanepoel & Avenant-Oldewage (1992) studied the glands associated with the proboscis and preoral spine of *A. japonicus* following the earlier research by Shimura & Inoue (1984). They reported that there were five glands and two gland cells associated with these mouthparts. These authors also described how glands one through four were linked by a single duct to the preoral spine and were positioned close to the dorsal body wall, while gland five and the two gland cells were associated with the proboscis and were positioned close to the proboscis and preoral spine (Swanepoel & Avenant-Oldewage, 1992). The year after, a study was conducted on the musculature associated with the suckers, preoral spine, and proboscis to show how each functions (Gresty et al., 1993). Watson & Avenant-Oldewage (1996) then conducted a brief, scanning electron microscopy study on the damage caused by *A. japonicus*, and showed that epidermal cells of the host fish, *Carassius auratus* (L.) were removed and that the proboscis caused damage. Reports then followed listing ulcerations and possible haemorrhagic changes to the host (Yildiz & Kumantas, 2002; Akter et al., 2007). Subsequently, a possible link was found between infections with *Argulus* species and the host's susceptibility to bacterial disease (Shimura et al., 1983b; Singhal et al., 1990; Bandilla et al., 2006). Bandilla et al. (2005) also proved that susceptibility of the fish host to infection by

Argulus species is not altered by a previous infection, and thus immunity to *Argulus* species infection is not acquired from a primary infection.

In their chapter on the Phylum Arthropoda, Lester & Hayward (2006) reported that the site of a wound inflicted by *Argulus* species appears as craters on low magnification formed by hyperplasia of the epidermis at the margins, and that these wounds may extend deep into the spongium compactum. In the same year, Tam & Avenant-Oldewage (2006) conducted a histological study on the larval digestive system of *Argulus japonicus*. In that study, they found that the larvae feed on mucus and epithelial cells only as their proboscis is not long enough to reach the dermal layer and their digestive system is not developed enough to digest complex food. Later, Forlenza et al. (2008) conducted a transcriptional analysis study of the common carp infected by *A. japonicus* larvae. Within this study changes in gene transcription in peripheral blood leucocytes and skin of carp infected by larval *A. japonicus* were examined. They showed that the response to infection is at first localised and restricted to the site of attachment, and thus feeding, but that over time it is extended throughout the skin as a whole organ if the infection is continued (Forlenza et al., 2008). In 2009, the ultrastructure of the digestive system of adult *A. japonicus* specimens was studied using transmission electron microscopy by Tam & Avenant-Oldewage (2009a). That study found that the digestive system of mature *A. japonicus* is similar to that of decapods and is able to digest lipids, amino acids, and glycogen which includes the blood of the host (Tam & Avenant-Oldewage, 2009a).

Research on *Argulus* species as intermediate hosts is relatively recent and thus only a few reports exist. Pfeil-Putzien (1977) and Pfeil-Putzien & Baath (1978) showed that Spring Viraemia of Carp (SVC) virus could be transmitted by *A. foliaceus*. Ahne (1985) then confirmed that *A. foliaceus* is the vector for the SVC virus. *Argulus coregoni* Thorell, 1864 has been linked as the intermediate host for skrjabillanid nematodes in Russia (Tikhomirova, 1970, 1980). Moravec (1978) showed that *A. foliaceus* is the intermediate host to *Molnaria erythrophthalmi* (Molnár, 1966), a skrjabillanid nematode, and Molnár & Moravec (1997) showed that second and third stage larvae nematodes are also present in *A. foliaceus*. Thereafter, Molnár & Székely (1998) drew the conclusion that the high prevalence of skrjabillanid nematodes in Hungary could be correlated to the high incidence of *A. foliaceus*, and Moravec et al. (1999) showed that *Argulus mexicanus* Pineda, Páramo & Del Rio, 1995 is the intermediate host for daniconematid nematodes in Mexico.

In a different approach, studies on the stress resulting from infection with *Argulus* species in combination with cortisol dosing of the hosts have been conducted by Ruane et al. (1999), van der Salm et al. (2000) and Haond et al. (2003). These studies showed that dosing of the food with cortisol at low levels will result in adaptive effects in the host but that high doses are maladaptive and result in failure of the host to elicit the characteristic stress response when subjected to infection with the parasite (Ruane et al., 1999; van der Salm et al., 2000; Haond et al., 2003). Although this may initially appear as a reduction in stress it is actually a negative feedback response from the host limited to the receptor cells in the epidermis (Ruane et al., 1999; van der Salm et al., 2000; Haond et al., 2003).

Treatment and control.— The need to control this parasite is evident from the effect that it has on its host. Apart from normal quarantine practices that are usually applied in the aquaculture industry, other physical, and chemical and lately an increase in biological methods have been suggested.

Physical control methods include practical, physical changes that can be made to the tanks to reduce the number of parasites present. Benz et al. (2001) suggested removing the parasites from the fish using forceps. Furthermore, egg laying plates in which the plates can be removed or replaced throughout the season so that the number of eggs that hatch in a fish farm is decreased was suggested by Hakalahti et al. (2004c). This method controls the numbers to eventually eliminate the population completely but, if a different substrate is used instead of the boards, the method becomes ineffective. This is also a time consuming practice. Recently, Parvez et al. (2013) used a similar mechanism but painted control substances on the wooden boards that were placed in the ponds. Three of the paints used were synthetic, namely "Byotrol" (Benzalkonium chloride), "Netrex" (copper (I) oxide) and "Chlorvar" (chlorinated rubber), and three were natural namely garlic, lime and neem (Parvez et al., 2013). The results of their 3 week investigation showed that wooden boards painted with Chlorvar were highly favourable to the *Argulus* spp. females to lay their eggs (Parvez et al., 2013). A shortcoming of the experiment was that Parvez et al. (2013) were uncertain if the eggs were killed quickly on the boards.

Chemical control methods involve the use of substances to disrupt the life cycle of the parasite and prevent generation. The chemicals used need to be effective for the argulid species but should have a negligible effect on the rest of the fauna and flora in the water system. The simplest chemical alternative would be to subject the fish to saltwater or freshwater dips so that they undergo osmotic shock (Benz et al., 2001). This would possibly dislodge the parasites but does not eliminate them from the water source. Puffer & Beal (1981) tested the use of 0,0-dimethyl 2,2,2-trichloro-1-hydroxyethylphosphonate (DTHP) in the treatment of branchiurans, however the DTHP acts as a cholinesterase inhibitor which affects the host fish in repetitive dosage (Puffer & Beal, 1981). Chowdhury et al. (2006) used an organophosphate pesticide, Sumithion 50 EC: Fenitrothion (0,0-dimethyl-0(3-methyl-nitrophenyl) phosphorothioate) on goldfish infected with *Argulus* spp. The results show that a 0.1 ppm solution of Sumithion 50 EC: Fenitrothion results in 100% mortality of argulids however, continual use of this pesticide resulted in negative effects on the fish as it is also a cholinesterase inhibitor, with low ovicidal properties.

The effect of Ivermectin and Doramectin as a feed, a bath and when injected was evaluated (Hemaprasanth et al., 2012). The results show that Ivermectin was more effective in all three methods but, the effects this drug would have on the fauna and flora of the water system still remains unknown. Emamectin benzoate (SLICE) has been tested both by Hakalahti et al. (2004b) and by Hanson et al. (2011). Hanson et al. (2011) found that a 0.2% emamectin benzoate solution administered in commercial fish feed was effective in a 7 day regimen with no fish mortalities.

The effect of pesticides on the environment prompted a recent trend to use natural extracts to control ectoparasites. An attempt at phytotherapy showed that the Long Pepper plant, *Piper longum* L., was effective at 9.0 mg/l over 48 h with no effective on the host *Carassius auratus* L. (see Kumar et al., 2012; Valladão et al., 2015). The use of aqueous extract from the leaf of *Azadirachta indica* A. Juss has also been explored by Kumar et al. (2013) and Banerjee et al. (2014b). Kumar et al. (2013) found that Azadirachtin administered in a bath treatment at 15 mg/l reduced haematological and biochemical parameters in the blood of goldfish; and Banerjee et al. (2014b) found that Azadirachtin interfered with the development of the oocytes in females of *Argulus bengalensis* Ramakrishna and thus could be used as an agent of control.

A recent trend in using the fish's immune response as a means of control has shown some promise as a biological control method. Ruane et al. (1995) provided the first account of a humoral response to *A. foliaceus* antigens in rainbow trout, *Oncorhynchus mykiss* (Walbaum) demonstrated by ELISA and immunoblotting. The study also showed cross reacting antigens in common with sea lice which, after further investigation, could result in broad-based vaccine against crustacean ectoparasites. In a similar line of enquiry, Saurabh et al. (2010) tested the modulation of the innate immune response of *Labeo rohita* (Hamilton) to infection with *A. siamensis*. The results of their study showed that an infection with *A. siamensis* causes the suppression of alpha-2 macroglobulin activity, serum complement activity and ceruplasmin levels through the induction of the stress response (Saurabh et al., 2010). These authors however indicated that further investigation is required to identify the mechanism of the fish innate immune response (Saurabh et al., 2010). The study into vaccines against *Argulus* spp. infection was further investigated by Sahoo et al. (2013). In their study, the whole transcriptome of *Argulus siamensis* Wilson, 1926 was analysed and characterised to understand the genetic makeup of the species. This study was the first to identify the genetics of an argulid species and can be used in future to create genetically targeted vaccines against argulosis (Sahoo et al., 2013). As techniques in genetics and biochemical studies improve, the use of biologically targeted mechanisms of control will become more common place.

Phylogeny of *Argulus* species

The phylogeny of *Argulus* species has received much attention for two reasons; the first is to understand the relationship of Branchiura to the Crustacea and other Arthropoda (see Branchiura section of this paper), and the second is to clarify the relationship to pentastomids (see Riley et al., 1978; Abele et al., 1989, 1992; Møller et al., 2008).

The earliest link made between Branchiura and the Pentastomida was by Wingstrand (1972) in his excellent study of sperm morphology in *A. foliaceus* and *Raillietiella hemidactyli* Hett, 1934. This was later supported by Riley et al. (1978), who used observations of the pentastomid embryo, integument, spermatogenesis and oogenises to highlight the similarilties between Crustacea and Pentastomida, and thus the branchiuran link to pentastomids. This was also confirmed with 18S ribosomal RNA by Abele et al. (1989); and Abele et al. (1992). This association between Branchiura and Pentastomida was later named the Ichthyostraca and found further support in other molecular studies (see Zrzavý, 2001; Edgecomb, 2010; Regier et al., 2010; Oakley et al., 2012).

The morphology of the maxillulae in *A. foliaceus* larvae and *D. ranarum* adults was compared and supplemented by the molecular phylogeny of *A. foliaceus*, *D. ranarum* and *Chonopeltis australis* Boxshall, 1976. These results were then compared to other maxillopodan representatives and members of the Pentastomida (see Møller et al., 2008) to position the genera of Branchiura to each other and within the maxillopodan group. The adult hooks of *Dolops* species was found to be homologous to larval hooks of *Argulus* species and furthermore, the ontogenetic sequence of development of maxillulae in *Argulus* species was found to recapitulate the phylogeny (Møller et al., 2008), confirming Haeckels' theory (Haeckel, 1866). The molecular data revealed three facts; the first is that the genus *Argulus* is paraphyletic with *C. australis*, the second is that *D. ranarum* is a sister group to the remaining Branchiura based on the retention of the hooks in the adult maxillulae; and the third is that they provided further support for a clade that includes Branchiura and Pentastomida as sister groups.

Dolops Audouin, 1837

The genus *Dolops* was erected by Audouin (1837) to represent the first species *Dolops lacordairei* Audouin, 1837 however, the species name was not followed by a description (Audouin, 1837) so the species has been declared "nomen nudum"

(Walter, 2015a). Thereafter, Heller (1857) established the genus *Gyropeltis* and named two of the species, *G. longicauda* Heller, 1857 and *G. kollari* Heller, 1857. Members of the genus *Gyropeltis* were identified as being similar to members of the genus *Argulus* but, they possess maxillulae modified as strong hooks instead of suckers, and lack a preoral spine (Heller, 1857). This genus name was later synonymised with *Dolops* by Bouvier (1899a). WoRMS lists *Gyropeltis* as an unaccepted synonym of *Dolops* (Walter, 2008).

Yamaguti (1963) listed the genus *Huargulus* Yu, 1938 with characters similar to the subfamily Dolopsinae Yamaguti, 1963 but Ku & Wang (1956) and later Fryer (1969) wrote that the description of this new species (*H. chinensis* Yu, 1938) was actually based on a juvenile *Argulus* specimen, which had been described by Tokioka in 1940 (see Fryer, 1969). The subfamily Dolopsinae was never accepted.

Synonyms of the genus Dolops Audouin, 1837

Dolops 1837
Gyropeltis Heller, 1857; *Dolops* see Bouvier (1899a).

Synonyms of the species of Dolops Audouin, 1837

Dolops bidentata (Bouvier, 1899)

Gyropeltis bidenata Bouvier, 1899
Dolops bidentata (Bouvier, 1899) according to Bouvier (1899a).

Dolops discoidalis Bouvier, 1899

Gyropeltis kollari Bouvier, 1897
Gyropeltis discoidalis Bouvier, 1899 according to Bouvier (1899a).

Dolops doradis Cornalia, 1860

Gyropeltis doradis Cornalia, 1860. Thorell (1864).
Dolops doradis Cornalia, 1860 according to Bouvier (1899b).

Dolops geayi (Bouvier, 1897)

Gyropeltis geayi Bouvier, 1897
Dolops geayi (Bouvier, 1897) according to Bouvier (1899a, b).

Dolops kollari (Heller, 1857)

Gyropeltis kollari Heller, 1857. Krøyer (1863); Thorell (1864).
Dolops kollari (Heller, 1857) according to Bouvier (1899a).

Dolops longicauda (Heller, 1857)

Gyropeltis longicauda Heller, 1857. Krøyer (1863); Thorell (1864).
Dolops longicauda (Heller, 1857) accourding to Bouvier (1899b).

Dolops ranarum (Stuhlmann, 1891)

Gyropeltis ranarum Stuhlmann, 1891
Dolops ranarum (Stuhlmann, 1891) according to Bouvier (1899b).

Dolops reperta (Bouvier, 1899)

Gyropeltis reperta Bouvier, 1899
Dolops reperta (Bouvier, 1899) according Bouvier (1899a).

Dolops striata (Bouvier, 1899)

Gyropeltis striata Bouvier, 1899
Dolops striata (Bouvier, 1899) according to Bouvier (1899a).

Dolops species and their distribution

WoRMS only recognises 13 *Dolops* species (Walter, 2015a). Most of these have been described from freshwater on the continent of South America (see table IX), but, two species have been found on other continents — *Dolops ranarum* (Stuhlmann, 1891) is exclusively found in Africa, and *D. tasmanianus* Fryer, 1969, has been found only in Tasmania. These two will be treated separately (tables X and XI).

Research conducted on *Dolops* species
Anatomy and physiology of *Dolops* species

Table XII lists references of publications on *Dolops* species.

Digestive system.— The first description of the digestive system was provided by Heller (1857). Much later, Avenant-Oldewage & Van As (1990a) used histological techniques to describe the digestive system as consisting of a pharynx, oesophagus, crop, enteral diverticula, midgut, and rectum. The blood meal taken by the parasite is also suggested to occur as a result of the mechanical damage caused primarily by the maxillular hooks (Avenant-Oldewage & Van As, 1990a). Recently, Tam & Avenant-Oldewage (2009b) used transmission electron microscopy to describe the ultrastructure of the digestive system of *D. ranarum*. The digestive system of *A. japonicus* and *D. ranarum* was found to be similar although they need to feed less frequesntly due to the storage of triglyceride lipids as a nutrient reserve (Tam & Avenant-Oldewage, 2009b).

TABLE IX
Freshwater *Dolops* spp. from the continent of South America

Dolops species	Country/city/town	Water system as published	Water system confirmed[a]	E	Host species	Host species confirmed[b]	Reference
Dolops bidentata (Bouvier, 1899)	Franco-bresilien Guyana	Riviere Lunier	Lunier River	F	Anguille	*Anguilla* sp.	Bouvier (1899a, b)
	Guiana	Uknown	Unknown	F	*Anguilla*	*Anguilla* sp.	Wilson (1902)
	Brazil	Lago Janauacá, Rio Solimões	Janauacá Lake, Solimões River	F	*Schizodon fasciatus*	*Schizodon fasciatus* Spix & Agassiz, 1829	Malta (1982a, 1984)
				F	*Rhytiodus microlepis*	*Rhytiodus microlepis* Kner, 1858	
				F	*Prochilodus nigricans*	*Prochilodus nigricans* Spix & Agassiz, 1829	
				F	*Colossoma bidens*	*Piaractus brachypomus* (Cuvier, 1818)	
				F	*Serrasalmus nattereri*	*Pygocentrus nattereri* Kner, 1858	
				F	*Astronatus ocellatus*	*Astronotus ocellatus* (Agassiz, 1831)	
	Pirizal district, Poconé Wetland, Mato Grosso, Brazil	Coqueiro Bay 16°15′12″S 56°22′12″W	Coqueiro Bay 16°15′12″S 56°22′12″W	F	*Pygocentrus nattereri*	*Pygocentrus nattereri* Kner, 1858	Silva-Souza et al. (2011)

TABLE IX
(Continued)

Dolops species	Country/city/town	Water system as published	Water system confirmed[a]	E	Host species	Host species confirmed[b]	Reference
	Northern Pantanal, Cáceres, Matto Grosso, Brazil	Caiçara Bay, Paraguay River between 16°05′02.8″S 57°44′22.7″W and 16°06′41.9″S 57°45′14.6″W	Caiçara Bay, Paraguay River between 16°05′02.8″S 57°44′22.7″W and 16°06′41.9″S 57°45′14.6″W	F	*Pygocentrus nattereri*	*Pygocentrus nattereri* Kner, 1858	Fontana et al. (2012)
				F	*Serrasalmus maculatus*	*Serrasalmus maculatus* Kner, 1858	
				F	*Serrasalmus marginatus* Valenciennes, 1847	*Serrasalmus marginatus* Valenciennes, 1837	
Dolops carvalhoi Lemos de Castro, 1949	Mato Grosso, Brazil	Rio Kuluene & Rio Xingu	Kuluene River-Xingu River Basin	F	*Rhaphiodon vulpinus*	*Rhaphiodon vulpinus* Spix & Agassiz, 1829	Lemos de Castro (1949)
	Brazil	Lago Janauacá	Janauacá Lake, Solimões River	F	*Pseudoplatystoma tigrinum*	*Pseudoplatystoma tigrinum* (Valenciennes, 1840)	Malta & Varella (1983); Malta (1984)
				F	*Pseudoplatystoma fasciatum*	*Pseudoplatystoma fasciatum* (Linnaeus, 1766)	
				F	*Phractocephalus hemiliopterus*	*Phractocephalus hemiliopterus* (Bloch & Schneider, 1801)	
				F	*Colossoma macropomum*	*Colossoma macropomum* (Cuvier, 1816)	

TABLE IX
(Continued)

Dolops species	Country/city/town	Water system as published	Water system confirmed[a]	E	Host species	Host species confirmed[b]	Reference
	Itacoatiara, Brazil	estação de piscicultura	fisheries station	F	*Rhaphiodon vulpinus*	*Rhaphiodon vulpinus* Spix & Agassiz, 1829	Gomes & Malta (2002)
				F	*Pellona castellneana*	*Pellona castelnaeana* Valenciennes, 1847	
				F	*Serrasalmus nattereri*	*Pygocentrus nattereri* Kner, 1858	
				F	*Prochilodus nigricans curimata*	*Prochilodus nigricans* Spix & Agassiz, 1829	
				F	*Cyprinus carpio*	*Cyprinus carpio* Linnaeus, 1758	
				F	*Colossoma macropomum* (Cuvier, 1818) tambaqui	*Colossoma macropomum* (Cuvier, 1816)	
	Miranda and Abobral, Pantanal	Miranda River Basin (Miranda, Vermelho & Abobral River, Baia da Medalha, Baia Negra, Baia Platina)	Miranda River Basin (Miranda, Vermelho & Abobral River, Baia da Medalha, Baia Negra, Baia Platina)	F	*Pygocentrus nattereri* Kner, 1860	*Pygocentrus nattereri* Kner, 1858	Carvalho et al. (2003)
				F	*Serrasalmus spilopleura* Kner, 1860	*Serrasalmus spilopleura* Kner, 1858	
				F	*Serrasalmus marginatus* Valenciennes, 1847	*Serrasalmus marginatus* Valenciennes, 1837	

TABLE IX
(Continued)

Dolops species	Country/city/town	Water system as published	Water system confirmed[a]	E	Host species	Host species confirmed[b]	Reference
	São José dos Bandeirantes, Nova Crixá, State of Goiás, Brazil	rio Araguaia basin, rio Araguaia (a tributary of the rio Tocantins) (13°41'S 50°47'W)	Araguaia River basin, Araguaia River (a tributary of the Tocantins River) (13°41'S 50°47'W)	F	*Pygocentrus nattereri* Kner, 1860	*Pygocentrus nattereri* Kner, 1858	Carvalho et al. (2004)
	Bolivia	Río Ichilo, oxbow lakes, and Río Beni	Ichilo River, oxbow lakes, and Beni River	F	*Pseudoplatystoma fasciatum* Linnaeus, 1840	*Pseudoplatystoma fasciatum* (Linnaeus, 1766)	Mamani et al. (2004)
				F	*Pseudoplatystoma tigrinum* Valenciennes, 1840	*Pseudoplatystoma tigrinum* (Valenciennes, 1840)	
	Brazil	floodplains of Upper Paraná River	Paraná River	F	*Pseudoplatystoma corruscans* (Spix & Agassiz, 1829)	*Pseudoplatystoma corruscans* (Spix & Agassiz, 1829)	Takemoto et al. (2009)
	South America (Imported to Japan)	Madeira River	Madeira River	F	*Asterophysus batrachus* Kner, 1857	*Asterophysus batrachus* Kner, 1858	Møller & Olesen (2012)
Dolops doradis Cornalia, 1860	Equatorial America	Not given	Not given	F	*Doras niger*	*Oxydoras niger* (Valenciennes, 1821)	Cornalia (1860)
	Central America	Unknown	Unknown	F	*Doras niger* Valenciennes	*Oxydoras niger* (Valenciennes, 1821)	Wilson (1902)

TABLE IX
(Continued)

Dolops species	Country/city/town	Water system as published	Water system confirmed[a]	E	Host species	Host species confirmed[b]	Reference
	Pirassununga, Sao Paulo, Brazil	rio Mogy-guassu	Mogy Guassú River	F	Salminus maxillosus C & V	Salminus brasiliensis (Cuvier, 1816)	Thomsen (1942)
				F	Salminus hilaris C & V	Salminus hilarii Valenciennes, 1850	
Dolops discoidalis Bouvier, 1899	Columbia	Sarare, Rio Nuba	Sarare River	F	Platysoma sp.	Platysoma sp.	Bouvier (1899a)
	Brazil	Rio Nuba	Sarare River	F	Platysoma sp. (Doncella)	Platysoma sp.	Wilson (1902)
	Matto-Grosso Brazil	Gy-Paraná River	Paraná River	F	Phractocephalus hemiliopterus Cast.	Phractocephalus hemioliopterus (Bloch & Schneider, 1801)	Moreira (1912, 1913)
	Taperinha, Amazon, Brazil	Unknown	Unknown	F	Arapaima gigas Cuvier	Arapaima gigas (Schinz, 1822)	Schuurmans Stekhoven (1937)
	Goiás, Brazil	Rio Cristalino	Cristalino River	F	Phractocephalus hemiliopterus Piraráras	Phractocephalus hemioliopterus (Bloch & Schneider, 1801)	Paiva Carvalho (1939)
	Rosario, Santa Fe, Argentina	Río Paraná	Paraná River	F	Hoplias malabaricus	Hoplias malabaricus (Bloch, 1794)	Ringuelet (1943)
	Edo. Bolivar, Venezuela	Rio Orinoco	Orinoco River	F	Pseudoplatystoma fasciatum (Linnaeus)	Pseudoplatystoma fasciatum (Linnaeus, 1766)	Weibezahn & Cobo (1964)

TABLE IX
(Continued)

Dolops species	Country/city/town	Water system as published	Water system confirmed[a]	E	Host species	Host species confirmed[b]	Reference
	Hata Santa Maria, Edo. Bolivar, Venezuela	Rio Orinoco	Orinoco River	F	Pseudoplatystoma fasciatum (Linnaeus)	Pseudoplatystoma fasciatum (Linnaeus, 1766)	
	Amazonas, Brazil	Alto Xingu	Alto Xingu Kuluene River	U	Host unknown	Host unknown	da Silva (1978)
	Brazil	Lago Janauacá, Rio Solimões	Janauacá Lake, Solimões River	F	Pseudoplatystoma fasciatum (Linnaeus)	Pseudoplatystoma fasciatum (Linnaeus, 1766)	Malta (1982a, 1984)
				F	Pseudoplatystoma tigrinum	Pseudoplatystoma tigrinum (Valenciennes, 1840)	
				F	Phractocephalus hemiliopterus	Phractocephalus hemioliopterus (Bloch & Schneider, 1801)	
				F	Leiarius marmoratus	Leiarius marmoratus (Gill, 1870)	
				F	Hemisorubim sp.	Hemisorubim platyrhynchos (Valenciennes, 1840)	
				F	Hoplerythrinus unitaeniatus	Hoplerythrinus unitaeniatus (Spix & Agassiz, 1829)	

TABLE IX
(Continued)

Dolops species	Country/city/town	Water system as published	Water system confirmed[a]	E	Host species	Host species confirmed[b]	Reference
				F	*Astronotus ocellatus*	*Astronotus ocellatus* (Agassiz, 1831)	
	Bolivia	Río Ichilo, oxbow lakes, and Río Beni	Ichilo River, oxbow lakes, and Beni River	F	*Arapaima gigas*	*Arapaima gigas* (Schinz, 1822)	Mamani et al. (2004)
				F	*Pseudoplatystoma fasciatum* Linnaeus, 1840	*Pseudoplatystoma fasciatum* (Linnaeus, 1766)	
				F	*Pseudoplatystoma tigrinum* Valenciennes, 1840	*Pseudoplatystoma tigrinum* (Valenciennes, 1840)	
Dolops geayi (Bouvier, 1897)	Venezuela	Lagoons between Apuré and Aranca	Lagoons between Apuré and Aranca	U	Not given	Not given	Bouvier (1897)
	Guiana	Apure and Aranca	Lagoons between Apuré and Aranca	U	Free swimming	Free swimming	Wilson (1902)
	Venezuela	Lake Velencia	Lake Velencia	F	*Aquidens pulcher* (Gill) Chusco	*Andinoacara pulcher* (Gill, 1858)	Pearse (1920)
				F	*Crenicichla geayi* Pellegrin, matuguaro	*Crenicichla geayi* Pellegrin, 1903	
				F	*Hoplias malabaricus* (Bloch) guabina	*Hoplias malabaricus* (Bloch, 1794)	

TABLE IX
(Continued)

Dolops species	Country/city/town	Water system as published	Water system confirmed[a]	E	Host species	Host species confirmed[b]	Reference
	Makthlawaiya	Paraguayan Chaco 58°19'W 23°25'S	Paraguayan Chaco 58°19'W 23°25'S, Paraguay River	F	*Hoplia malabaricus* (Bloch)	*Hoplias malabaricus* (Bloch, 1794)	Cunnington (1931)
	Rosario, Santa Fe, Argentina	Río Paraná	Paraná River	F	*Salminus maxillosus*	*Salminus brasiliensis* (Cuvier, 1816)	Ringuelet (1943, 1948)
				F	*Hoplias malabaricus*	*Hoplias malabaricus* (Bloch, 1794)	
	Venezuela	Cano Guariquito, Edo. Guarico	Cano Guariquito stream	F	*Phractocephalus hemiliopterus* (Bloch & Schneider)	*Phractocephalus hemiliopterus* (Bloch & Schneider, 1801)	Weibezahn & Cobo (1964)
	Brazil	Lago Janauacá	Janauacá Lake, Solimões River	F	*Megalodoras* sp.	*Megalodoras* sp.	Malta (1982b, 1984)
				F	*Crenicichla* sp.	*Crenicichla* sp.	
				F	*Hoplias malabaricus*	*Hoplias malabaricus* (Bloch, 1794)	
	Venezuala	Cano Guariquito, Edo. Guarico	Cano Guariquito stream	F	*Astronotus ocellatus*	*Astronotus ocellatus* (Agassiz, 1831)	Malta (1984)
	Brazil	Lago Janauacá, Rio Solimões	Janauacá Lake, Solimões River	F	*Megalodoras* sp.	*Megalodoras* sp.	Malta (1984)

TABLE IX
(Continued)

Dolops species	Country/city/town	Water system as published	Water system confirmed[a]	E	Host species	Host species confirmed[b]	Reference
	Brazil	Floodplains of Upper Paraná River	Paraná River	F	*Prochilodus lineatus* (Valenciennes, 1837)	*Prochilodus lineatus* (Valenciennes, 1837)	Takemoto et al. (2009)
Dolops intermedia da Silva, 1978	Santo Antonio da Patrulha, Rio Grande do Sul, Brazil		Gravataí River	F	*Hoplias malabaricus*	*Hoplias malabaricus* (Bloch, 1794)	da Silva (1978)
	Arroio Teixeira, Rio Grande do Sul, Brazil		South Atlantic Ocean	B	*Crenicichla* sp.	*Crenicichla* sp.	
	Pantano Grande, Rio Grande do Sul, Brazil		Jacui River	F	*Hoplias malabaricus*	*Hoplias malabaricus* (Bloch, 1794)	
	Guaiba, Rio Grande do Sul, Brazil		Guaiba Lake	U	Host unknown	Host unknown	
Dolops kollari (Heller, 1857)	Brazil	Unknown	Unknown	U	Host unknown	Host unknown	Heller (1857)
	Venzuela	Rio Nuba	Sarare River	F	*Platystoma* sp.	*Platysoma* sp.	Bouvier (1897)
	Brazil	Unknown	Unknown	U	Host unknown	Host unknown	Wilson (1902)
	Brazil	Unknown	Unknown	U	Host unknown	Host unknown	Moreira (1912, 1913)
Dolops lacordairei Audouin, 1837[c]	Cayenne	Not given	Not given	F	Aymara	*Hoplerythrinus unitaeniatus* (Spix & Agassiz, 1829)	Audouin (1837)

TABLE IX
(Continued)

Dolops species	Country/city/town	Water system as published	Water system confirmed[a]	E	Host species	Host species confirmed[b]	Reference
Dolops longicauda (Heller, 1857)	Brazil	Not given	Not given	F	Hydrocyon brevidens	Salminus brasiliensis (Cuvier, 1816)	Heller (1857)
	Brazil	Not given	Not given	F	Hydrocyon brevidens	Salminus brasiliensis (Cuvier, 1816)	Krøyer (1863)
	Brazil	Unknown	Unknown	F	Hydrocyon brevidens	Salminus brasiliensis (Cuvier, 1816)	Wilson (1902)
	São Paulo, Brazil	Not given	Not given	F	Salminus brevidens Cuv.	Salminus franciscanus Lima & Britski, 2007	Maidl (1912)
	Caceres, Matto-Grosso, Brazil	Paraguay River	Paraguay River	F	Salminus brevidens (Cuv. & Val.) dorado	Salminus franciscanus Lima & Britski, 2007	Moreira (1912, 1913)
	S. Manoel & S. Paulo, Brazil	Rio Araquá	Araquá River	F	Dourado	Salminus brasiliensis (Cuvier, 1816)	Paiva Carvalho (1939)
	Argentina	Cataratas del Iguazú	Iguazu Falls	U	Host unknown	Host unknown	Ringuelet (1943)
	Anegadizos, Argentina	Río Paraná	Paraná River	U	Host unknown	Host unknown	
	Rosario, Santa Fe, Argentina	Río Paraná	Paraná River	F	Salminus maxillosus	Salminus brasiliensis (Cuvier, 1816)	

TABLE IX
(Continued)

Dolops species	Country/city/town	Water system as published	Water system confirmed[a]	E	Host species	Host species confirmed[b]	Reference
				F	Serrasalmus nattereri	Pygocentrus nattereri Kner, 1858	Brian (1947)
	Ribereña, Argentina	Río de la Plata	Río de la Plata	F	Salminus maxillosus	Salminus brasiliensis (Cuvier, 1816)	
	Corpus, Argentina	Río Paraná	Paraná River	F	Salminus maxillosus	Salminus brasiliensis (Cuvier, 1816)	
	Entre Rios	Arroyo Las Conchas, Paraná	Las Conchas stream	F	Salminus maxillosus	Salminus brasiliensis (Cuvier, 1816)	
	Argentina	Rio Gualeguaychu	Gualeguaychú River	F	Salminus brevidens (Cuv. & Val.) dorado	Salminus franciscanus Lima & Britski, 2007	
	Pto. Basilio Argentina	Rio Uruguay	Uruguay River	F	Salminus brevidens (Cuv. & Val.) dorado	Salminus franciscanus Lima & Britski, 2007	
	Zarate, Argentina	Rio Parana	Paraná River	F	Salminus brevidens (Cuv. & Val.) dorado	Salminus franciscanus Lima & Britski, 2007	
	San Pedro, Buenos Aires, Argentina	Rio Parana	Paraná River	U	Host unknown	Host unknown	
	San Fernando, Argentina	Rio Lujan	Luján River	F	Salminus brevidens (Cuv. & Val.) dorado	Salminus franciscanus Lima & Britski, 2007	

TABLE IX
(Continued)

Dolops species	Country/city/town	Water system as published	Water system confirmed[a]	E	Host species	Host species confirmed[b]	Reference
	Rosario, Santa Fe, Argentina	Río Paraná	Paraná River	F	Pseudoplatystoma coruscans	Pseudoplatystoma coruscans (Spix & Agassiz, 1829)	Ringuelet (1948)
	Rosario, Santa Fe, Argentina	Río Paraná	Paraná River	F	Salminus nattereri	Pygocentrus nattereri Kner, 1858	
	Entre Rios, Argentina	Río Gualeguaychú	Gualeguaychú River	F	Salminus maxillosus	Salminus brasiliensis (Cuvier, 1816)	
	Santa Fe, Argentina	Río Paraná	Paraná River	F	Potamotrygon sp.	Potamotrygon sp.	
	Entre Rios, Argentina	Ibicuycito	Ibicuycito	F	Salminus maxillosus	Salminus brasiliensis (Cuvier, 1816)	
	Entre Rios, Argentina	Sauces de Merlo, Gualeguaychú	Gualeguaychú River	F	Salminus maxillosus	Salminus brasiliensis (Cuvier, 1816)	
	Santa Fe, Argentina	Río Colastiné	Colastiné River, tributary of the Paraná River	F	Salminus maxillosus	Salminus brasiliensis (Cuvier, 1816)	
	Misiones, Argentina	Alto Uruguay	Uruguay River	F	Salminus maxillosus	Salminus brasiliensis (Cuvier, 1816)	
	Ecuador	Limoncocha Lake	Limoncocha Lake	F	Serrasalmus sp.	Serrasalmus sp.	Hugghins (1970)

TABLE IX
(Continued)

Dolops species	Country/city/town	Water system as published	Water system confirmed[a]	E	Host species	Host species confirmed[b]	Reference
				F	*Aequidens tetramerus* (Heckel)	*Aequidens tetramerus* (Heckel, 1840)	
	Brazil	floodplains of Upper Paraná River	Paraná River	F	*Salminus brasiliensis* (Cuvier, 1816)	*Salminus brasiliensis* (Cuvier, 1816)	Takemoto et al. (2009)
Dolops nana Lemos de Castro, 1950	Minas Gerais, Brazil	Ribeiro do Itaci, Carmo do Rio Claro	Sapucaí River	F	"dourado" (*Salminus* sp.)	*Salminus brasiliensis* (Cuvier, 1816)	Lemos de Castro (1950)
	Brazil	Río Paraná 22°50′22°70′S 53°15′53°40′W	Paraná River 22°50′22°70′S 53°15′53°40′W	F	*Leporinus friderici* (Bloch, 1794)	*Leporinus friderici* (Bloch, 1794)	Guidelli et al. (2006)
	Brazil	floodplains of Upper Paraná River	Paraná River	F	*Leporinus friderici* (Bloch, 1794)	*Leporinus friderici* (Bloch, 1794)	Takemoto et al. (2009)
				F	*Leporinus elongatus* Valenciennes, 1850	*Leporinus elongatus* Valenciennes, 1850	
				F	*Leprinus obtusidens* (Valenciennes, 1837)	*Leporinus obtusidens* (Valenciennes, 1837)	
Dolops reperta (Bouvier, 1899)	Contesté franco-brésilien Guyana	Rivière Lunier	Lunier River	F	Aymara	*Hoplerythrinus unitaeniatus* (Spix & Agassiz, 1829)	Bouvier (1899a, b); Wilson (1902)
Dolops striata (Bouvier, 1899)	Guyana	Unknown	Unknown	B	Anguille	*Anguilla* sp.	Bouvier (1899a, c)

TABLE IX
(Continued)

Dolops species	Country/city/town	Water system as published	Water system confirmed[a]	E	Host species	Host species confirmed[b]	Reference
	Guyana Makthlawaiya	Unknown	Unknown	B	Anguilla	Anguilla sp.	Wilson (1902)
		Paraguayan Chaco 58°19'W 23°25'S	Paraguayan Chaco 58°19'W 23°25'S, Paraguay River	F	Hoplias malabaricus (Bloch)	Hoplias malabaricus (Bloch, 1794)	Cunnington (1931)
	Entre Rios, Argentina	La Brea, Gualeguaychú	La Brea, Gualeguaychú River	F	Hoplias malabaricus	Hoplias malabaricus (Bloch, 1794)	Ringuelet (1948)
	Misiones, Argentina	Pindapoy	Pindapoy stream	F	Salminus maxillosus	Salminus brasiliensis (Cuvier, 1816)	
	Rosario, Santa Fe, Argentina	Río Paraná	Paraná River	F	Hoplias malabaricus	Hoplias malabaricus (Bloch, 1794)	
	Argentina	Laguna Iberá	Iberá wetlands	F	Cynodon vulpinus Spix	Rhaphiodon vulpinus Spix & Agassiz, 1829	Schuurmans Stekhoven (1951)
	Edo. Guarico, Venezuela	Rio Guarico	Guárico River	F	Hoplias malabaricus (Bloch)	Hoplias malabaricus (Bloch, 1794)	Weibezahn & Cobo (1964)
	Edo. Guarico, Venezuela	Laguna La Tigra, Cano Guariquito	Laguna La Tigra, Cano Guariquito stream	F	Hoplias malabaricus (Bloch)	Hoplias malabaricus (Bloch, 1794)	
	Ecuador	Limoncocha Lake	Limoncocha Lake	F	Serrasalmus sp.	Serrasalmus sp.	Hugghins (1970)

TABLE IX
(Continued)

Dolops species	Country/city/town	Water system as published	Water system confirmed[a]	E	Host species	E	Host species confirmed[b]	Reference
	Bolivia	Tamichucua Lake	Tamichucua Lake	F	*Hoplias malabaricus* (Bloch)	F	*Hoplias malabaricus* (Bloch, 1794)	Malta & Varella (1983); Malta (1984)
	Brazil	Lago Janauacá, Rio Solimões	Janauacá Lake, Solimões River	F	*Schizodon fasciatus*	F	*Schizodon fasciatus* Spix & Agassiz, 1829	
				F	*Leporinus fasciatus*	F	*Leporinus fasciatus* (Bloch, 1794)	
				F	*Leporinus* sp.	F	*Leporinus* sp.	

E, environment in which it is found: B, brackish water; F, freshwater; M, marine; U, unknown.

[a]) Water system confirmed.

[b]) Host species confirmed using fishbase.org.

[c]) *Dolops lacordairei* Audouin, 1837 has been declared "nomen nudum" because of the lack of a species description (Audouin, 1837; Bouvier, 1899a; Yamaguti, 1963; Walter, 2015a).

TABLE X
African distribution records for *Dolops ranarum* (Stuhlmann, 1891)

Country/city/town	Water system as published	Water system confirmed[a]	E	Host species	Host species confirmed[b]	Reference
Buboka, Western Nyansa, Africa	Not given	Lake Malawi	E	Frog tadpoles	Frog tadpoles	Stuhlmann (1891)
Tanzania	Kala	Lake Tanganyika	F	*Lates microlepsis*	*Lates microlepis* Boulenger, 1898	Cunnington (1913)
	Lake Victoria, Nyanza	Lake Victoria	F	*Bagrus degeni* Nfui	*Bagrus degeni* Boulenger, 1906	
			F	*Protopterus ethiopicus*	*Protopterus aethiopicus* Heckel, 1851	
			F	*Clarias angillaris*	*Clarias anguillaris* (Linnaeus, 1758)	
	Lake Nyasa	Lake Malawi	F	Sungwa	*Serranochromis robustus* (Günther, 1864)	
	Lake No, White Nile	Lake No, White Nile	F	*Heterobranchus bidorsalis* Hala	*Heterobranchus bidorsalis* Geoffroy Saint-Hilaire, 1809	
Ozeguru, Democratic Republic of Congo	Lake Albert	Lake Albert	F	Not given	Not given	Brian (1940)
Pretoria, South Africa	Hartebeest Poort Dam	Hartebeestpoort Dam	F	*Tilapia mossambica*	*Oreochromis mossambicus* (Peters, 1852)	Barnard (1955)
Pretoria, South Africa	Apies River	Apies River	F	*Barbus gunningi*	*Labeobarbus marequensis* (Smith, 1841)	
Middleburg, South Africa	Olifants River	Olifants River	F	*Barbus swierstrae*	*Labeobarbus marequensis* (Smith, 1841)	
Bulawayo, Zimbabwe	Matopo Dam	Matopo Dam	F	*Huro salmonoides* (Black Bass)	*Micropterus salmoides* (Lacépède, 1802)	
	River Banga, a tributary of the Luweya	Lake Malawi	F	*Clarias mossambicus* Peters	*Clarias gariepinus* (Burchell, 1822)	Fryer (1956)

TABLE X
(Continued)

Country/city/town	Water system as published	Water system confirmed[a]	E	Host species	Host species confirmed[b]	Reference
	Nkata Bay, Lake Nyasa	Lake Malawi	F	*Clarias mossambicus* Peters	*Clarias gariepinus* (Burchell, 1822)	Fryer (1959)
	Bana Lagoon	Bana Lagoon	F	*Clarias mossambicus* Peters	*Clarias gariepinus* (Burchell, 1822)	
Zimbabwe	Lake Bangweulu	Lake Bangweulu	F	*Chrysichthys*	*Chrysichthys* sp.	
			F	*Entropius*	*Entropius* sp.	
			F	*Schilbe*	*Schilbe* sp.	
			F	*Clarias*	*Clarias* sp.	
			F	*Heterobranchus*	*Heterobranchus*	
Leopoldville	River Congo	Congo River	F	*Ophiocephalus obscurus* Günther	*Parachanna obscura* (Günther, 1861)	Fryer (1960a)
Chavuma, near Balovale, Zimbabwe	North Kasisi (or Kashizki) River, Upper Zambezi	Upper Zambezi	F	*Clarias mossambicus* Peters	*Clarias gariepinus* (Burchell, 1822)	
Chiboboma	Zambezi River	Zambezi River	F	*Tilapia mossambica* (Peters)	*Oreochromis mossambicus* (Peters, 1852)	
Mazabuka	Kafue River	Kafue River	F	*Serranochromis* sp.	*Serranochromis* sp.	
Fort Johnstone	Lake Nyasa	Lake Malawi	F	Host unknown	Host unknown	
Uganda	Lake Nabugabo	Lake Nabugabo	F	Not recorded	Not recorded	
Kenya	Lake Baringo	Lake Baringo	F	Not recorded	Not recorded	
Northern Rhodesia	Lake Bangweulu	Lake Bangweulu	F	Not recorded	Not recorded	Fryer (1960b)
unkown	Lake Victoria	Lake Victoria	F	Not recorded	Not recorded	
	Lake Victoria	Lake Victoria	F	*Protopterus aethiopicus* Heckel	*Protopterus aethiopicus* Heckel 1851	Fryer (1961a)

TABLE X
(Continued)

Country/city/town	Water system as published	Water system confirmed[a]	E	Host species	Host species confirmed[b]	Reference
unkown	Lake Victoria	Lake Victoria	F	Clarias mossambicus Peters	Clarias gariepinus (Burchell, 1822)	
unkown	Lake Victoria	Lake Victoria	F	Bagrus doemac Forskal	Bagrus docmak (Forskål, 1775)	
unkown	Lake Victoria	Lake Victoria	F	Tilapia variabilis Boulenger	Oreochromis variabilis (Boulenger, 1906)	
unkown	Lake Victoria	Lake Victoria	F	Tilapia esculenta Graham	Oreochromis esculentus (Graham, 1928)	
Diafarabé, Senegal	River Niger 14°16'N 5°5'W	Niger River 14°16'N 5°5'W	F	Tetraodon fahaka L.	Tetraodon lineatus Linnaeus, 1758	Fryer (1964)
Vila Perreira de Eça, Angola	swamp 17°S 15°40'E	swamp 17°S 15°40'E	F	Not recorded	Not recorded	
Uganda	Lake George	Lake George	F	Bagrus doemac Forskal	Bagrus docmak (Forskål, 1775)	
unkown	Lake Mweru	Lake Mweru	F	Labeo altivelis Peters	Labeo altivelis Peters, 1852	Fryer (1965a)
unkown	Lake Mweru	Lake Mweru	F	Tilapia macrochir Boulenger	Oreochromis macrochir (Boulenger, 1912)	
Ethiopia	Lake Tana	Lake Tana	F	Tilapia nilotica L.	Oreochromis niloticus (Linnaeus, 1758)	Fryer (1965b)
Ethiopia	Lake Tana	Lake Tana	F	Varichorhinus sp.	Varichorhinus sp.	
Ethiopia	Koka Dam	Koka Reservoir	F	Tilapia sp.	Tilapia sp.	
unkown	Lake Albert	Lake Albert	F	Auchenoglanis occidentalis (Cuvier & Valenciennes)	Auchenoglanis occidentalis (Valenciennes, 1840)	
unkown	Lake Albert	Lake Albert	F	Bagrus bajad (Forskal)	Bagrus bajad (Forskål, 1775)	

TABLE X
(Continued)

Country/city/town	Water system as published	Water system confirmed[a]	E	Host species	Host species confirmed[b]	Reference
unkown	Lake Albert	Lake Albert	F	*Bagrus doemac* Forskal	*Bagrus docmak* (Forsskål, 1775)	
unkown	Lake Albert	Lake Albert	F	*Clarias lazera* Cuvier & Valenciennes	*Clarias gariepinus* (Burchell, 1822)	
unkown	Lake Albert	Lake Albert	F	*Lates niloticus* (L.)	*Lates niloticus* (Linnaeus, 1758)	Thurston (1970)
Uganda	Lake Victoria	Lake Victoria	F	*Bagrus doemac*	*Bagrus docmak* (Forsskål, 1775)	
Uganda	Lake Victoria	Lake Victoria	F	*Clarias mossambicus*	*Clarias gariepinus* (Burchell, 1822)	
Uganda	Lake Victoria	Lake Victoria	F	*Astatoreochromis alluandi*	*Astatoreochromis alluaudi* Pellegrin (1904)	
Uganda	Lake Albert	Lake Albert	F	*Lates albertianus*	*Lates niloticus* (Linnaeus, 1758)	Thurston (1970)
Uganda	Lake Kyoga	Lake Kyoga	F	*Lates albertianus*	*Lates niloticus* (Linnaeus, 1758)	
Uganda	Victoria Nile	Nile River	F	*Lates albertianus*	*Lates niloticus* (Linnaeus, 1758)	
Uganda	Lake Victoria	Lake Victoria	F	*Bagrus docmac* Forskahl	*Bagrus docmak* (Forsskål, 1775)	Mbahinzireki (1980)
South Africa	Doorndraai Dam	Doorndraai Dam	F	*Barbus marequensis*	*Labeobarbus marequensis* (Smith, 1841)	Avenant & Van As (1985)
			F	*Oreochromis mossambicus*	*Oreochromis mossambicus* (Peters, 1852)	
	Fanie Botha Dam	Tzaneen Dam	F	*Barbus marequensis*	*Labeobarbus marequensis* (Smith, 1841)	
			F	*Clarias gariepinus*	*Clarias gariepinus* (Burchell, 1822)	
			F	*Micropterus dolomieu*	*Micropterus dolomieu* Lacépède, 1802	
			F	*Oreochromis mossambicus*	*Oreochromis mossambicus* (Peters, 1852)	

TABLE X
(Continued)

Country/city/town	Water system as published	Water system confirmed[a]	E	Host species	Host species confirmed[b]	Reference
	Glen Alpine Dam	Glen Alpine Dam	F	*Oreochromis mossambicus*	*Oreochromis mossambicus* (Peters, 1852)	
			F	*Barbus marequensis*	*Labeobarbus marequensis* (Smith, 1841)	
			F	*Labeo rubropunctatus*	*Labeo congoro* Peters, 1852	
			F	*Eutropius depressirostris*	*Schilbe mystus* (Linnaeus, 1758)	
			F	*Clarias gariepinus*	*Clarias gariepinus* (Burchell, 1822)	
	Luphephe and Nwanedzi Dam	Luphephe and Nwanedi Dam	F	*Barbus marequensis*	*Labeobarbus marequensis* (Smith, 1841)	
			F	*Eutropius depressirostris*	*Schilbe mystus* (Linnaeus, 1758)	
			F	*Clarias gariepinus*	*Clarias gariepinus* (Burchell, 1822)	
	Nzhelele Dam	Nzhelele Dam	F	*Oreochromis mossambicus*	*Oreochromis mossambicus* (Peters, 1852)	
			F	*Barbus marequensis*	*Labeobarbus marequensis* (Smith, 1841)	
			F	*Clarias gariepinus*	*Clarias gariepinus* (Burchell, 1822)	
	Roodeplaat Dam 25 40′S 28 20′E	Roodeplaat Dam 25 40′S 28 20′E	F	*Oreochromis mossambicus*	*Oreochromis mossambicus* (Peters, 1852)	
			F	*Barbus mattozi*	*Barbus mattozi* Guimarães, 1884	

TABLE X
(Continued)

Country/city/town	Water system as published	Water system confirmed[a]	E	Host species	Host species confirmed[b]	Reference
			F	*Clarias gariepinus*	*Clarias gariepinus* (Burchell, 1822)	Avenant & Van As (1985, 1986)
			F	*Oreochromis mossambicus*	*Oreochromis mossambicus* (Peters, 1852)	
			F	*Chetia flaviventris*	*Chetia flaviventris* Trewavas, 1961	Avenant & Van As (1985)
Northern KwaZulu Natal	Pongola flood plains	Pongolapoort Dam	F	Not recorded	Not recorded	Avenant et al. (1989a)
Northern Transvaal	Limpopo Drainage system	Limpopo Drainage system	F	Not recorded	Not recorded	
Republic of Venda, Southern Africa	Luphephe and Nwanedzi Dam	Luphephe and Nwanedi Dam	F	*Barbus marequensis*	*Labeobarbus marequensis* (Smith, 1841)	Avenant et al. (1989b)
			F	*Eutropius depressirostris*	*Schilbe mystus* (Linnaeus, 1758)	
			F	*Clarias gariepinus*	*Clarias gariepinus* (Burchell, 1822)	
			F	*Oreochromis mossambicus*	*Oreochromis mossambicus* (Peters, 1852)	
Northern Transvaal	Limpopo Drainage system	Limpopo Drainage system	F	*Clarias gariepinus*	*Clarias gariepinus* (Burchell, 1822)	Avenant-Oldewage (1994b)
Zimbabwe	Lake Kariba	Lake Kariba	F	*Oreochromis mortimeri* (Trewavas, 1966) Kariba Tilapia	*Oreochromis mortimeri* (Trewavas, 1966)	Douëllou & Erlwanger (1994)

TABLE X
(Continued)

Country/city/town	Water system as published	Water system confirmed[a]	E	Host species	F	Host species confirmed[b]	Reference
Zimbabwe	Lake Kariba	Lake Kariba			F	Pseudocrenilabrus philander (Weber, 1897) Southern Mouthbrooder	Pseudocrenilabrus philander (Weber, 1897)	
Zimbabwe	Lake Kariba	Lake Kariba			F	Serranochromis codringtonii (Boulenger, 1908) Green Happy	Sargochromis codringtonii (Boulenger, 1908)	
Zimbabwe	Lake Kariba	Lake Kariba			F	Serranochromis macrocephalus (Boulenger, 1899) Purpleface Largemouth	Serranochromis macrocephalus (Boulenger, 1899)	
Zimbabwe	Lake Kariba	Lake Kariba			F	Tilapia rendalli rendalli (Boulenger, 1906) Redbreast Tilapia	Tilapia rendalli (Boulenger, 1897)	
Zimbabwe	Lake Kariba	Lake Kariba			F	Clarias gariepinus (Burchell, 1822) Sharptooth Catfish	Clarias gariepinus (Burchell, 1822)	
Zimbabwe	Lake Kariba	Lake Kariba			F	Synodontis zambezensis Peters, 1852 Brown Squeaker	Synodontis zambezensis Peters, 1852	
Zimbabwe	Lake Kariba	Lake Kariba			F	Mormyrops deliciosus (Leach, 1818) Cornish Jack	Mormyrops anguilloides (Linnaeus, 1758)	

TABLE X
(Continued)

Country/city/town	Water system as published	Water system confirmed[a]	E	F	Host species	Host species confirmed[b]	Reference
Zimbabwe	Lake Kariba	Lake Kariba		F	*Mormyrus longirostris* Peters, 1852 Eastern Bottlenose	*Mormyrus longirostris* Peters, 1852	Møller et al. (2008)
Tzaneen, Limpopo Province, South Africa	Fanie Botha Dam	Tzaneen Dam		F	*Clarias gariepinus* (Burchell, 1822) Sharptooth Catfish	*Clarias gariepinus* (Burchell, 1822)	Tam & Avenant-Oldewage (2009b)
Limpopo Province, South Africa	Luphephe and Nwanedi Dam	Luphephe and Nwanedi Dam		F	*Clarias gariepinus* (Burchell, 1822) Sharptooth Catfish	*Clarias gariepinus* (Burchell, 1822)	Tam & Avenant-Oldewage (2009b)
Limpopo Province, South Africa	Luphephe and Nwanedi Dam	Luphephe and Nwanedi Dam		F	*Oreochromis mossambicus* (Peters, 1852)	*Oreochromis mossambicus* (Peters, 1852)	Van As & Van As (2015)
Botswana	Okavango River and Delta	Okavango River and Delta		F	*Hepsetus odoe*	*Hepsetus odoe* (Bloch, 1794)	
				F	*Schilbe intermedius*	*Schilbe intermedius* Rüppell, 1832	
				F	*Clarias gariepinus*	*Clarias gariepinus* (Burchell, 1822)	
				F	*Clarias stappersii*	*Clarias stappersii* Boulenger, 1915	
				F	*Synodontis nigromaculatus*	*Synodontis nigromaculatus* Boulenger, 1905	
				F	*Oreochromis andersonii*	*Oreochromis andersonii* (Castelnau, 1861)	
				F	*Oreochromis macrochir*	*Oreochromis macrochir* (Boulenger, 1912)	

TABLE X
(Continued)

Country/city/town	Water system as published	Water system confirmed[a]	E	Host species	Host species confirmed[b]	Reference
			F	*Sargochromis carlottae*	*Sargochromis carlottae* (Boulenger, 1905)	
			F	*Sargochromis giardia*	*Sargochromis giardi* (Pellegrin, 1903)	
			F	*Serranochromis robustus*	*Serranochromis robustus* (Günther, 1864)	
			F	*Serranochromis macrocephalus*	*Serranochromis macrocephalus* (Boulenger, 1899)	
			F	*Coptodon rendalli*	*Tilapia zillii* (Gervais, 1848)	

E, environment in which it is found: B, brackish water; F, freshwater; M, marine; U, unknown.
[a] Water system confirmed.
[b] Host species confirmed using fishbase.org.

TABLE XI

Australian distribution records for *Dolops tasmanianus* Fryer, 1969

Country/city/town	Water system as published	Water system confirmed[a]	E	Host species	F	host species confirmed[b]	Reference
Tasmania	Lake Surprise	Lake Surprise		*Galaxias* sp.		*Galaxias* sp.	Fryer (1969)

E, environment in which it is found: F, freshwater.

[a]) Water system confirmed.

[b]) Host species confirmed using fishbase.org.

TABLE XII

Summary of selected records of the studies on the anatomy and physiology conducted on *Dolops* species

	Species	References
Digestive system	*Dolops* spp.	Heller (1857); Cornalia (1860)
	D. longicauda	Maidl (1912)
	D. ranarum	Avenant-Oldewage & Van As (1990a); Avenant-Oldewage (1994b); Tam & Avenant-Oldewage (2009b)
Nervous system	*Dolops* spp.	Heller (1857); Cornalia (1860)
	D. longicauda	Maidl (1912)
Sensory system	*Dolops* spp.	Heller (1857)
Respiratory system	*Dolops* spp.	Heller (1857)
Circulatory system	*Dolops* spp.	Heller (1857); Cornalia (1860)
	D. longicauda	Maidl (1912)
Reproductive system male	*Dolops* spp.	Heller (1857); Cornalia (1860)
	D. ranarum	Fryer (1958, 1960b)
	D. geayi	Fryer (1958)
Reproductive system female	*Dolops* spp.	Heller (1857); Cornalia (1860)
	D. longicauda	Maidl (1912)
	D. ranarum	Avenant-Oldewage & Van As (1990b)
Life cycle	*D. carvalhoi*	Gomes & Malta (2002); Møller & Olesen (2012)
	D. ranarum	Fryer (1964); Avenant et al. (1989b)

Nervous system.— The last description of the nervous system of *Dolops* species was provided by Maidl (1912). In this description, Maidl (1912) describes the nervous system as consisting of two optic nerves that originate from the cerebral ganglion and the suboesophageal ganglion. This is by contrast to Martin (1932), Zaćwilichowska (1948) and Madsen (1964) who later described one pair of cerebral ganglia, one pair of suboesophageal ganglia and five pairs of ventral ganglia in *Argulus*. The nervous system of *Dolops* species may need to be re-examined in order to make a more accurate comparison to the nervous system of *Argulus* species. A similar re-examination needs to occur for the sensory and respiratory system of *Dolops* species as these have not been studied since Heller (1857) and Maidl (1912).

Reproductive system male.— The male reproductive system of *D. ranarum* was described in great detail by Fryer (1960b). The reproductive system is similar to that of *Argulus* species except for the tri-lobed testes which *D. ranarum* possesses (Fryer, 1960b; Neethling & Avenant-Oldewage, 2015). Fryer (1960b) was also the first researcher to report the occurrence of spermatophores in branchiurans.

Reproductive system female.— The reproductive system of female *D. ranarum* has been described by Avenant-Oldewage & Van As (1990b) and was found to be similar to that of *Argulus* species.

Life cycle.— Fryer (1964) was the first to describe the larva of *D. ranarum* as being a juvenile adult because it hatches with the maxillulae as a single hook, as in the adult form, instead of the two hooks on the maxillulae in *Argulus* species larvae. This description was confirmed by Avenant et al. (1989b). Recently, Møller & Olesen (2012) described the first stage larva of *D. carvalhoi* Lemos de Castro, 1949 and compared it to the descriptions of *D. ranarum* by Fryer (1964) and Avenant et al. (1989b). These authors found only slight differences in the morphology of the larvae which could be attributed to the difference in species (Møller & Olesen, 2012). It was suggested that studies on other species of *Dolops* may contribute to the knowledge of their origins as evidence now points to the genus originating in South America (see table IX), but the existence of species in South Africa and Tasmania could either mean a remnant of Gondwanian origin or an introduction to these two continents (Fryer, 1969; Møller & Olesen, 2012).

Parasite-host interactions in *Dolops* species (fig. 5F-G)

Effect on host.— The effect that *Dolops* species have on their host is directly linked to their attachment, and secondarily to feeding (Avenant-Oldewage & Van As, 1990a). Avenant-Oldewage (1994b) found that the attachment by the maxillulae hooks causes haemorrhaging from the dermal layer of the skin and swelling from oedema. Secondary infection by bacteria and excessive mucous production was also found. In a recent study, Tavares-Dias et al. (2007) monitored the changes in blood parameters of fish parasitized by *D. carvalhoi*. They found that parasitized fish blood had a lower haematocrit but increased plasma glucose, serum protein sodium and serum protein chloride levels when compared to the control fish; while the blood smears showed an increase in monocytes and PAS-positve granular leukocytes. These changes were attributed to *D. carvalhoi* feeding on the host blood (Tavares-Dias et al., 2007). The combination of the effects found by both Avenant-Oldewage (1994b) and Tavares-Dias et al. (2007) has implications for fish farmers in that a severe outbreak can occur in high density stocks and will decrease the ability of the fish to resist other immune challenges that may be presented, as was found in *Argulus* species.

Chonopeltis Thiele, 1900

The genus *Chonopeltis* Thiele, 1900 was established to accommodate the first species *C. inermis* Thiele, 1900. Thiele (1900) characterised members of this genus as having antennae with four segments, maxillulae modified as suckers and supporting rods in the head shield. The maxillae and swimming legs were described as similar to members of the genus *Argulus* (see Thiele, 1900). Members of this genus lack antennulae and a preoral spine (Thiele, 1900).

Synonyms of the species of *Chonopeltis* Thiele, 1900
Chonopeltis australissimus Fryer, 1977

Chonopeltis minutus Fryer, 1977 according to Van As & Van As (1999a).

Chonopeltis species and their distribution

This genus is entirely endemic to Africa and WoRMS only recognises 15 *Chonopeltis* species with one subspecies (Walter, 2015b). A compendium of the distribution of 13 of these species was provided by Avenant-Oldewage & Knight (1994) relaying relative measurements, location, site and host information, as well as drawings of diagnostic characteristics used for each of the species (Avenant-Oldewage & Knight, 1994). Van As & Van As (1999a) have since synonymised *C. australissimus* Fryer, 1977 with *C. minutus*, but this has not been recognised by WoRMS (Walter, 2015b). The current distribution table (table XIII) therefore has information for all the species.

Research conducted on *Chonopeltis* species

Unlike the genera *Argulus* and *Dolops*, biological research on the genus *Chonopeltis* is limited to only a few aspects of anatomy and physiology.

Anatomy and physiology of *Chonopeltis* species

Digestive system.— Studies on the digestive system of *Chonopeltis* species are limited to two studies on *C. australis* Boxshall, 1976. The first by Swanepoel & Avenant-Oldewage (1993) described the foregut of *C. australis* using histology and live staining of specimens. Their study showed that the foregut is divided into a preoral cavity, an ascending oesophagus and a horizontal oesophagus that protrudes into the anterior midgut. From their study, Swanepoel & Avenant-Oldewage (1993) suggested that *Chonopeltis* spp. feed on mucus and cellular detritus. The next year Avenant-Oldewage et al. (1994) provided a complete description of the digestive system of the *C. australis* that included the foregut, anterior and posterior midgut, and the hindgut. They also discussed the terminology used to describe the digestive tract of branchiurans by various authors.

Reproductive system male.— The male reproductive system has recently been described by Neethling & Avenant-Oldewage (2015). They also described the occurrence of a spermatophore to transfer sperm in the genus for the first time. Before this, research on the reproductive system of *Chonopeltis* species had been restricted to the use of the accessory copulatory structures as a means of identification by Van As & Van As (1999b). The female reproductive system has yet to be described.

TABLE XIII

Selected records of African freshwater locations from which *Chonopeltis* species were recorded

Chonopeltis species	Country/city/town	Water system as published	Water system confirmed[a]	E	Host species	Host species confirmed[b]	Reference
Chonopeltis australis Boxshall, 1976	Vereeniging, South Africa	Vaal River	Vaal River	F	*Labeo capensis* Smith	*Labeo capensis* (Smith, 1841)	Boxshall (1976)
	Potchefstroom, South Africa	Boskop Dam reservoir, Mooi River	Boskop Dam reservoir, Mooi River	F	*Labeo rosae* Steindachner	*Labeo rosae* Steindachner, 1894	
	Bloemfontein, South Africa	Wuras Dam 29°40'S 26°00'E	Wuras Dam 29°40'S 26°00'E	F	*Barbus aeneus* (Hocitt & Skelton, 1983)	*Labeobarbus aeneus* (Burchell, 1822)	Van Niekerk & Kok (1989)
	Bloemfontein, South Africa			F	*Labeo capensis* Smith, 1841	*Labeo capensis* (Smith, 1841)	
	South Africa	Orange-Vaal River System	Orange-Vaal River System	F	*Labeo capensis* (Smith, 1841)	*Labeo capensis* (Smith, 1841)	Van As & Van As (1999a)
	Potchefstroom, South Africa	Boskop Dam	Boskop Dam reservoir, Mooi River	F	*Labeo umbratus*	*Labeo umbratus* (Smith, 1841)	Knight & Avenant-Oldewage (1990)
	Potchefstroom, South Africa	Boskop Dam	Boskop Dam reservoir, Mooi River	F	*Labeo umbratus* (Smith, 1841)	*Labeo umbratus* (Smith, 1841)	Swanepoel & Avenant-Oldewage (1993)
	Potchefstroom, South Africa			F	*Labeo capensis* (Smith, 1841)	*Labeo capensis* (Smith, 1841)	
	Potchefstroom, South Africa	Boskop Dam	Boskop Dam reservoir, Mooi River	F	*Labeo umbratus* (Smith, 1841)	*Labeo umbratus* (Smith, 1841)	Avenant-Oldewage & Knight (1994)
				F	*Labeo capensis*	*Labeo capensis* (Smith, 1841)	

TABLE XIII
(Continued)

Chonopeltis species	Country/city/town	Water system as published	Water system confirmed[a]	E	Host species	Host species confirmed[b]	Reference
	Potchefstroom, South Africa	Boskop Dam	Boskop Dam reservoir, Mooi River	F	Labeo umbratus (Smith, 1841)	Labeo umbratus (Smith, 1841)	Avenant-Oldewage et al. (1994)
				F	Labeo capensis (Smith, 1841)	Labeo capensis (Smith, 1841)	
	Potchefstroom, South Africa	Boskop Dam 27°08'E	Boskop Dam reservoir 27°08'E	F	Labeo umbratus (Smith, 1841)	Labeo umbratus (Smith, 1841)	Avenant-Oldewage & Knight (2008)
	Potchefstroom, South Africa	26°33'S, Mooi River, a tributary of the Vaal River	26°33'S, Mooi River	F	Labeo capensis (Smith, 1841)	Labeo capensis (Smith, 1841)	
	Bloemfontein, South Africa	Maselspoort Dam	Maselspoort Dam, Modder River	F	Host unknown	Host unknown	Møller et al. (2008)
	South Africa	Vaal River 26°49'14.73"S 28°3'52.49"E	Vaal River 26°49'14.73"S 28°3'52.49"E	F	Labeo capensis (Smith, 1841)	Labeo capensis (Smith, 1841)	Neethling & Avenant-Oldewage (2015)
Chonopeltis australissimus Fryer, 1977[c]	Western Cape, South Africa	Great Berg River	Great Berg River	F	Barbus burgi Boulenger	Pseudobarbus burgi (Boulenger, 1911)	Fryer (1977); Van As & Van As (1999a)
Chonopeltis brevis Fryer, 1961	Not given	Victoria Nile	Nile River	F	Barbus altianalis radcliffi Boulenger	Barbus altianalis Boulenger, 1900	Fryer (1961a)
	Not given	Lake Victoria	Lake Victoria	F	Labeo victorianus Boulenger	Labeo victorianus Boulenger, 1901	

TABLE XIII
(Continued)

Chonopeltis species	Country/city/town	Water system as published	Water system confirmed[a]	E	Host species	Host species confirmed[b]	Reference
	Kenya	Ragati River	Ragati River	F	Amphilius grandis Boulenger	Amphilius grandis Boulenger, 1905	Fryer (1961b)
	Kenya	Ragati River	Ragati River	F	Garra sp.	Garra sp.	
	Tanganyika	Mugambuzi River	Mugambuzi River	F	Amphilius sp.	Amphilius sp.	
	Merilla, Kenya	Tana River	Tana River	F	Labeo cylindricus Peters	Labeo cylindricus Peters, 1852	Fryer (1964)
	Sagana, Kenya	Regati River (tributary of Tana River)	Ragati River	F	Labeo cylindricus Peters	Labeo cylindricus Peters, 1852	
	Nigeria	Cross River Estuary	Cross River Estuary	F	Chrysichthys nigrodigitatus	Chrysichthys nigrodigitatus (Lacépède, 1803)	Obiekezie et al. (1988)
Chonopeltis congicus Fryer, 1959[d]	Zimbabwe	Lake Bangweulu	Lake Bangweulu	F	Gnathonemus monteiri	Marcusenius monteiri (Günther, 1873)	Fryer (1959)
Chonopeltis elongatus Fryer, 1974	Kindu, Zaire	Lualaba River	Lualaba River	F	Synodontis longirostris Boulenger	Synodontis longirostris Boulenger, 1902	Fryer (1974)
Chonopeltis flaccifrons Fryer, 1960a	Congo	Lake Mweru	Lake Mweru	F	Marcusenius discorhynchus (Peters)	Cyphomyrus discorhynchus (Peters, 1852)	Fryer (1960a)
	M'Bunzi	River Fimi	Fimi River	F	Marcusenius welwerthi Boulenger	Hippopotamyrus wilverthi (Boulenger, 1898)	
	unknown	Malagarasi Swamps	Malagarasi River Swamps	F	Marcusenius sp.	Marcusenius sp.	

TABLE XIII
(Continued)

Chonopeltis species	Country/city/town	Water system as published	Water system confirmed[a]	E	Host species	Host species confirmed[b]	Reference
Chonopeltis fryeri Van As, 1986	Limpopo province (Northern Transvaal)	Mogalakwena River 23°20'S 28°40'E	Mogalakwena River 23°20'S 28°40'E	F	Clarias theodorae Weber, 1897	Clarias theodorae Weber, 1897	Van As (1986)
	Potchefstroom, South Africa	Loskop Dam, Olifants River	Loskop Dam, Olifants River	F	Clarias gariepinus	Clarias gariepinus (Burchell, 1822)	
Chonopeltis inermis Thiele, 1900	East Africa Wiedenhafen	A port	Indian Ocean	M	Chromis sp.	Chromis sp.	Thiele (1900)
	Wiedenhafen		Indian Ocean	M	Chromis sp.	Chromis sp.	Wilson (1902)
	Nkata Bay District	River Banga	Lake Malawi	F	Clarias mossambicus Peters	Clarias gariepinus (Burchell, 1822)	Fryer (1956)
Chonopeltis inermis schoutedenii Brian, 1940	Ango-Ango, Democratic Republic of Congo	Congo River	Congo River	F	Not given	Not given	Brian (1940)
	Mushie, Democratic Republic of Congo	Kasai River	Kasai River	F	Gnathonemus schilthuisiae	Marcusenius schilthuisiae (Boulenger, 1899)	Dartevelle (1951)
				F	Gnathonemus stanleyanus	Marcusenius stanleyanus (Boulenger, 1897)	
	Mbunzi, Democratic Republic of Congo	Fimi River	Fimi River	F	Marcusenius wilverthi	Hippopotamyrus wilverthi (Boulenger, 1898)	
	Botswana	Okawango River	Okavango River	F	Synodontis melanostictus	Synodontis melanostictus Boulenger, 1906	Barnard (1955)

TABLE XIII
(Continued)

Chonopeltis species	Country/city/town	Water system as published	Water system confirmed[a]	E	F	Host species	Host species confirmed[b]	Reference
	Angola	Diolo Lake	Diolo Lake		F	*Mormyrus anchiaetae* Guimarães, 1884	*Mormyrus lacerda* Castelnau, 1861	Marques (1978)
	South Africa	Luphephe River 22°38′S 30°24′E	Luphephe River 22°38′S 30°24′E		F	*Clarias theodorae* Weber, 1897	*Clarias theodorae* Weber, 1897	Van As & Van As (1993)
	Malawi	Lake Malawi	Lake Malawi		F	*Bathyclarias nyasensis* (Worthington, 1933)	*Bathyclarias nyasensis* (Worthington, 1933)	
Chonopeltis koki Van As, 1992	Katima Mulilo, Namibia	Zambezi River 17°10′S 24°15′E	Zambezi River 17°10′S 24°15′E		F	*Labeo cylindricus* Peters, 1852	*Labeo cylindricus* Peters, 1852	Van As (1992)
Chonopeltis lisikili Van As & Van As, 1996	Okavango	Thomalakane River, Okavango Delta 19°45′S 23°30′E	Thomalakane River, Okavango Delta 19°45′S 23°30′E		F	*Synodontis leopardinus* Pellegrin (1914)	*Synodontis leopardinus* Pellegrin, 1914	Van As & Van As (1996)
	Botswana	Okavango River and Delta	Okavango River and Delta		F	*Synodontis leopardinus*	*Synodontis leopardinus* Pellegrin, 1914	Van As & Van As (2015)
						Synodontis macrostigma	*Synodontis macrostigma* Boulenger, 1911	
						Synodontis nigromaculatus	*Synodontis nigromaculatus* Boulenger, 1905	

TABLE XIII
(Continued)

Chonopeltis species	Country/city/town	Water system as published	Water system confirmed[a]	E	Host species	F	Host species confirmed[b]	Reference
					Synodontis thamalakanensis		*Synodontis thamalakanensis* Fowler, 1935	Van As & Van As (1999c)
					Synodontis vanderwaali		*Synodontis vanderwaali* Skelton & White, 1990	
Chonopeltis liversedgi Van As & Van As, 1999	Okavango	Boro River swamps 19°26'S 22°49'E	Boro River swamps 19°26'S 22°49'E		*Mormyrus lacerda* Castelnau, 1861	F	*Mormyrus lacerda* Castelnau, 1861	
	Okavango	Okavango River lagoon 18°23'S 21°51'E	Okavango River lagoon 18°23'S 21°51'E		*Mormyrus lacerda* Castelnau, 1861	F	*Mormyrus lacerda* Castelnau, 1861	
	Okavango	Kalatog Channel 18°25'S 21°56'E	Kalatog Channel 18°25'S 21°56'E		*Mormyrus lacerda* Castelnau, 1861	F	*Mormyrus lacerda* Castelnau, 1861	
	Botswana	Okavango River and Delta	Okavango River and Delta		*Mormyrus lacerda* Castelnau, 1861	F	*Mormyrus lacerda* Castelnau, 1861	Van As & Van As (2015)
Chonopeltis meridionalis Fryer, 1964	Zimbabwe	Nuanetzi River	Mwenezi River		*Labeo rosae* Steindachner	F	*Labeo rosae* Steindachner, 1894	Fryer (1964)
Chonopeltis minutus Fryer, 1977	Western Cape, South Africa	Twee and Tra Tra Rivers of the Olifants system	Twee and Tra Tra Rivers of the Olifants system (Cape Province)		*Barbus calidus* Barnard	F	*Barbus calidus* Barnard, 1938	Fryer (1977); Van As & Van As (1999a)

TABLE XIII
(Continued)

Chonopeltis species	Country/city/town	Water system as published	Water system confirmed[a]	E	Host species	F	Host species confirmed[b]	Reference
	Western Cape, South Africa	Twee and Tra Tra Rivers of the Olifants system	Twee and Tra Tra Rivers of the Olifants system (Cape Province)	F	*Barbus erubescens* Skelton		*Barbus erubescens* Skelton (1974)	Fryer (1977); Van As & Van As (1999a)
Chonopeltis schoutedeni Brian, 1940[e]	Zimbabwe	Lake Bangweulu	Lake Bangweulu	F	*Gnathonemus* spp. (*G. monteiri*)		*Marcusenius monteiri* (Günther, 1873)	Fryer (1959)
				F	*Mormyrus longirostris* Peters		*Mormyrus longirostris* Peters, 1852	
	Iragalla	Malagarasi Swamps	Malagarasi River Swamps	F	*Mormyrus* sp.		*Mormyrus* sp.	Fryer (1960a)
	Not given	Lake Mweru	Lake Mweru	F	*Marcusenius discorhynchus* (Peters)		*Cyphomyrus discorhynchus* (Peters, 1852)	
	Not given	Lake Mweru	Lake Mweru	F	*Gnathonemus moeruensis* Boulenger		*Marcusenius macrolepidotus* (Peters, 1852)	Fryer (1965a)
Chonopeltis victori Avenant-Oldewage, 1991	Kruger National Park	Olifants River 31°45'E 24°59'S	Olifants River 31°45'E 24°59'S	F	*Labeo rosae* Steindachner, 1894		*Labeo rosae* Steindachner, 1894	Avenant-Oldewage (1991)
	Kruger National Park	Olifants River 31°45'E 24°59'S	Olifants River 31°45'E 24°59'S	F	*Labeo cylindricus*		*Labeo cylindricus* Peters, 1852	Avenant-Oldewage & Knight (1994)
	Kruger National Park	Olifants River	Olifants River	F	*Labeo rosae* Steindachner, 1894		*Labeo rosae* Steindachner, 1894	Luus-Powell & Avenant-Oldewage (1996)

TABLE XIII
(Continued)

Chonopeltis species	Country/city/town	Water system as published	Water system confirmed[a]	E	Host species	Host species confirmed[b]	Reference
	Kruger National Park	Olifants River	Olifants River	F	Labeo congoro Peters, 1852 (= syn. L. rubropunctatus)	Labeo congoro Peters, 1852	
	Kruger National Park	Olifants River	Olifants River	F	Labeo ruddi Boulenger (1907)	Labeo ruddi Boulenger, 1907	
	Kruger National Park	Olifants River	Olifants River	F	Labeo cylindricus Peters, 1852	Labeo cylindricus Peters, 1852	Luus-Powell & Avenant-Oldewage (1996)
	Kruger National Park	Olifants River	Olifants River	F	Barbus marequensis Smith, 1841	Labeobarbus marequensis (Smith, 1841)	Luus-Powell & Avenant-Oldewage (1996)

E, environment in which it is found: B, brackish water; F, freshwater; M, marine; U, unknown.

[a]) Water system confirmed.

[b]) Host species confirmed using fishbase.org.

[c]) Junior synonym of C. minutus according to Van As & Van As (1999a).

[d]) Was C. inermis v. schoutedeni Brian & C. inermis v. schoutedeni Dertevelle according to Fryer (1959).

[e]) Was C. inermis v. schoutedeni Brian & C. inermis v. schoutedeni Dertevelle according to Fryer (1959).

Life cycle.— Barnard (1955) briefly described the changes that occur in the maxillulae of three juvenile *Chonopeltis inermis* specimens kept at the museum. Thereafter, Fryer (1956) provided a detailed description of the first larval stage of *C. inermis*. Fryer (1961b) then compared the larval development of *C. inermis* to *C. brevis* Fryer, 1961 but, provided descriptions only for stages of *C. brevis* he had managed to collect. Fryer (1977) later described development in *C. minutus* and concluded that they are similar to *C. brevis*. The first scanning electron microscopy study of larvae was done on *C. australis* (see van Niekerk & Kok, 1989). Most recently, Van As & Van As (1996) described larval development of *C. lisikili* Van As & Van As, 1996 and found that development is similar to that of *C. brevis* but that it retains rudiments of larval hooks on fully developed maxillulae suckers (Van As & Van As, 1996). The life cycle descriptions still lack a description of all the stages.

Dipteropeltis Calman, 1912

Members of the fourth genus, *Dipteropeltis* Calman, 1912 are described as members of Argulidae with maxillae modified as suckers, a preoral spine that lacks a spine, simple antennae and antennulae, the lack of abdominal furcal rami, and a cephalon that bears elongated carapace lobes (Calman, 1912). This was based on a single species, *Dipteropeltis hirundo* Calman, 1912.

Similarly, Moreira (1912) wrote a brief description of a new genus and species that he called *Talaus ribeiroi* Moreira, 1912; the description given differed only slightly from the description of *D. hirundo* Calman, 1912. The genus *Talaus* was then renamed *Moreiriella* by Melo-Leitão as the genus *Talaus* was already in use in the class Arachnida by Simon (Moreira, 1915). Both *Talaus* and *Moreiriella* were synonymised with *Dipteropeltis* by Moreira (1915) who cited that they were "of the same place of origin". *Talaus (Moreiriella) ribeiroi* was thus synonymised with *D. hirundo* and *Talaus* became a junior homonym. Since these, a new species, *Dipteropeltis campanaformis* Neethling, Malta & Avenant-Oldewage, 2014 has been described.

Based on this description, the characteristics of *Dipteropeltis* were modified slightly so that the diagnosis now reads: Argulidae with maxillulae modified as suckers that lack supporting rods, a preoral spine may be present, simple antennae and antennulae, the presence or absence of abdominal furcal rami, and a cephalon that bears elongated carapace lobes.

Synonyms of the genus *Dipteropeltis* Calman, 1912

Dipteropeltis Calman, 1912
Talaus Moreira, 1912; *Dipteropeltis* see Moreira (1915).
Moreiriella Melo-Leitão, 1914; *Dipteropeltis* see Moreira (1915).

TABLE XIV

Selected records of freshwater *Dipteropeltis* species collected from South America

Dipteropeltis species	Country/city/town	Water system as published	Water system confirmed[a]	E	Host species	Host species confirmed[b]	Reference
Dipteropeltis campanaformis Neethling, Malta & Avenant-Oldewage, 2014	Manaus, Amazonas, Brazil	"igarapé", Reserva Florestal Ducke	Unnamed forest stream ("igarapé", Água Branco)	F	*Brycon amazonicus* (Spix & Agassiz, 1829)	*Brycon amazonicus* (Spix & Agassiz, 1829)	Neethling et al. (2014)
Dipteropeltis hirundo Calman, 1912	Amazonas, Brazil	Padauari River	Padauari River	F	*Acestrorhynchus* sp.	*Acestrorhynchus* sp.	Calman (1912); Møller & Olesen (2010)
	Corumbá, Matto Grosso, Brazil	Not given	Paraguay River	F	"Dourado"	*Salminus brasiliensis* (Cuvier, 1816)	
	Boa Esperança, São Paulo	Rio Jacaré-guassú	Jacaré Pepira River	F	*Tetragonopterus rutilus*	*Astyanax fasciatus* (Cuvier, 1819)	Paiva Carvalho (1941)
	Rosario, Argentina	Río Paraná medio	Perana River	F	*Salminus maxillosus*	*Salminus brasiliensis* (Cuvier, 1816)	Ringuelet (1943, 1948)
				F	*Luciopimelodus pati*	*Luciopimelodus pati* (Valenciennes, 1835)	
	Guárico state, Venezuela	Río Tiznado	Tiznado River	F	*Brycon whitei*	*Brycon whitei* Myers & Weitzman, 1960	Weibezahn & Cobo (1964)
	Guárico state, Venezuela	Caño Guariquito	Caño Guariquito intermittent stream	F	*Serrasalmus spilopleura*	*Serrasalmus spilopleura* Kner, 1858	

TABLE XIV
(Continued)

Dipteropeltis species	Country/city/town	Water system as published	Water system confirmed[a]	E	Host species	Host species confirmed[b]	Reference
	Miranda and Abobral, Pantanal	Abobral River	Abobral River	F	Pygocentrus nattereri	Pygocentrus nattereri Kner, 1858	Carvalho et al. (2003)
	Bolivia	Río Ichilo basin	Ichilo River basin	F	Pseudoplatystoma fasciatum	Pseudoplatystoma fasciatum (Linnaeus, 1766)	Mamani et al. (2004)

E, environment in which it is found: F, freshwater.
a) Water system confirmed.
b) Host species confirmed using fishbase.org.

Synonyms of *Dipteropeltis hirundo* Calman, 1912

Talaus ribeiroi Moreira, 1913; *Dipteropeltis hirundo* see Moreira (1915).
Moreiriella ribeiroi Moreira, 1915; *Dipteropeltis hirundo* see Moreira (1915).

Dipteropeltis species and their distribution

This species is endemic to the Amazon and WoRMS currently only lists a single species of *Dipteropeltis* Calman, 1912 (Boxshall & Walter, 2015). A second species, *Dipteropeltis campanaformis* Neethling, Malta & Avenant-Oldewage (2014), has recently been described. The current distribution table (table XIV) therefore has information for both species.

Research conducted on *Dipteropeltis* species

The literature on *Dipteropeltis* species is limited to species descriptions and distribution records, with only females documented.

ACKNOWLEDGEMENTS

We wish to extend our gratitude to the National Research Foundation and the University of Johannesburg for funding and the University for providing infrastructure. Dr. Ole Møller and an unknown reviewer are thanked for their extensive input on the first version of the manuscript.

APPENDIX

Classification of Branchiura:
According to Linnaeus, 1758
Kingdom Animalia
   Class Insecta
      Order Aptera
         Genus *Monoculus* Linnaeus, 1758

According to Latreille, 1802
Kingdom Animalia
   Class Crustacea
      Subclass Entomostraca
         Order Pseudopoda Latreille, 1802
            Genus *Argulus* Müller, 1785

According to Leach, 1819
Kingdom Animalia
   Class Crustacea
      Subclass Entomostraca
         Order Pæcilopoda Latreille, 1802
            Family Argulidae Leach, 1819
               Genus *Argulus* Müller, 1785

According to H. Milne Edwards, 1834
Kingdom Animalia
   Class Crustacea

Subclass Suçeur
  Legion 1 Parasites nageurs
    Order Siphonostoma
      Family Argulidae Leach, 1819
        Genus *Argulus* Müller, 1785

According to H. Milne Edwards, 1840
Kingdom Animalia
  Class Crustacea
    Subclass Suçeur
      Legion 1 Peltocephales
      Tribe Arguliens
        Order Siphonostoma Latreille, 1802
          Family Argulidae Leach, 1819
            Genus *Argulus* Müller, 1785

According to Baird, 1850
Kingdom Animalia
  Subkingdom Annulosa
    Class Crustacea
      Division Entomostraca
        Legion 1 Pæcilopoda
          Order Siphonostoma
            Tribe Peltocephala
              Family Argulidae Leach, 1819
                Genus *Argulus* Müller, 1785

According to Zenker, 1854
1. Malacostraca.
Trilobita 2. 7. Ostracoda.
  Poecilopoda 3. 5. Entomostraca. 6. Cirrhipedia.
(Argulina a.) 4. Aspidostraca (b. Branchiopoda).

According to Thorell, 1864
Kingdom Animalia
  Class Branchiopoda
    Order Branchiura Thorell, 1864
      Family Argulidae Leach, 1819
        Genus *Argulus* Müller, 1785
        Genus *Gyropeltis* Heller, 1857

According to Yamaguti, 1963
Subclass Branchiura Thorell, 1864
  Order Argulidea Yamaguti, 1963
    Family Argulidae Leach, 1819
      Subfamily Argulinae Yamaguti, 1963
        Genus *Argulus* Müller, 1785
      Subfamily Chonopeltinae Yamaguti, 1963
        Genus *Chonopeltis* Thiele, 1900
      Subfamily Dolopsinae Yamaguti, 1963
        Genus *Dolops* Audouin, 1837
        Genus *Huargulus* Yü, 1938
    Family Dipteropeltidae Yamaguti, 1963

Genus *Dipteropeltis* Calman, 1912
Genus *Talaus* Moreira, 1912

REFERENCES

ABELE, L. G., W. KIM & B. E. FELGENHAUER, 1989. Molecular evidence for inclusion of the Phylum Pentastomida in the Crustacea. Molecular Biology and Evolution, **6**(6): 685-691.

ABELE, L. G., T. SPEARS, W. KIM & M. APPLEGATE, 1992. Phylogeny of selected maxillopodan and other crustacean taxa based on 18S ribosomal nucleotide sequences: a preliminary analysis. Acta Zoologica (Stockholm), **73**(5): 373-382.

AHNE, W., 1985. *Argulus foliaceus* L. and *Piscicola geometra* L. as mechanical vectors of spring viraemia of carp virus (SVCV). Journal of Fish Diseases, **8**: 241-242.

AKTER, M. A., M. D. HOSSAIN & M. R. RAHMAN, 2007. Parasitic diseases of exotic carp in Bangladesh. Journal of Agriculture and Rural Development, **5**(1-2): 127-134.

ALAŞ, A., A. ÖKTENER & K. SOLAK, 2010. A study of the morphology of *Argulus foliaceus* Lin., 1758 (Crustacea; Branchiura) procured from Çavuşcu Lake (central Anatolia — Turkey) with scanning electron microscopy. Turkish Journal of Biology, **34**: 147-151.

ALI, N. M., E. S. ABDUL-EIS, K. N. ABDUL-AMEER & L. S. KADIM, 1988. On the occurrence of fish parasites raised in man-made lakes 1 — Protozoa and Crustacea. Journal of Biological Sciences Research, **19**(Suppl.): 877-885.

ALLUM, M. O. & E. J. HUGGHINS, 1959. Epizootics of fish lice, *Argulus biromosus*, in two lakes of eastern South Dakota. Journal of Parasitology, **45**(4): 33-34.

ALSARAKIBI, M., H. WADEH & G. LI, 2014. Influence of environmental factors on *Argulus japonicus* occurrence of Guangdong province, China. Parasitology Research, **113**(11): 4073-4083.

AMIN, O. M., 1981. On the crustacean ectoparasites of fishes from southeast Wisconsin. Transactions of the American Microscopical Society, **100**(2): 142-150.

ARÉVALO, C., 1921. Un Nuevo *Argulus* español. Boletin de la Sociedad Española de Historia Natural, **21**: 108-110. [In Spanish.]

ASENCIO, G., D. ROGEL, J. CARVAJAL & M. T. GONZALEZ, 2010. Life cycle of *Argulus araucanus* on Chilean rock cod *Eleginops maclovinus* from Valdivia River, southern Chile. 8th International Sea Lice Conference, Victoria, British Columbia, Canada: 10.

ATRIA, G. G., 1975. Una nueva especie die *Argulus*: *A. araucanus* n. sp. Boletin. Museo Nacional de Historia Natural (Chile), **34**: 81-85. [In Spanish.]

AUDOUIN, M., 1837. Annales de la Société Entomologique de France, **6**: 13. [In French.]

AVENANT, A., G. C. LOOTS & J. G. VAN AS, 1989a. A redescription of *Dolops ranarum* (Stuhlmann, 1891) (Crustacea: Branchiura). Systematic Parasitology, **13**: 141-151. DOI:10. 1007/BF00015223.

AVENANT, A. & J. G. VAN AS, 1985. Occurrence and distribution of *Dolops ranarum* Stuhlmann, ectoparasite of freshwater fish in the Transvaal, south Africa. Journal of Fish Biology, **27**: 403-416.

AVENANT, A. & J. G. VAN AS, 1986. Observations on the seasonal occurrence of the fish ectoparasite *Dolops ranarum* (Stuhlmann, 1891) (Crustacea: Branchiura) in the Transvaal. South African Journal of Wildlife Research, **16**(2): 62-64.

AVENANT, A., J. G. VAN AS & G. C. LOOTS, 1989b. On the hatching and morphology of *Dolops ranarum* larvae (Crustacea: Branchiura). Journal of Zoology: Proceedings of the Zoological Society of London, **217**: 511-519.

AVENANT-OLDEWAGE, A., 1991. A new species of *Chonopeltis* (Crustacea: Branchiura) from the Kruger National Park, southern Africa. Journal of African Zoology, **105**: 313-321.

AVENANT-OLDEWAGE, A., 1994a. A new species of *Argulus* from Kosi Bay, South Africa and distribution records of the genus. Koedoe, **37**(2): 89-95.

AVENANT-OLDEWAGE, A., 1994b. Integumental damage caused by *Dolops ranarum* (Stuhlmann, 1891) (Crustacea: Branchiura) to *Clarias gariepinus* (Burchell), with reference to normal histology and wound-inflicting structures. Journal of Fish Diseases, **17**: 641-647.

AVENANT-OLDEWAGE, A., 2001. *Argulus japonicus* in the Olifants River system-possible conservation threat? South African Journal of Wildlife Research, **31**(1&2): 59-63.

AVENANT-OLDEWAGE, A. & L. EVERTS, 2010. *Argulus japonicus*: sperm transfer by means of a spermatophore on *Carrasius auratus* (L.). Experimental Parasitology, **126**: 232-238.

AVENANT-OLDEWAGE, A. & E. KNIGHT, 1994. A diagnostic species compendium of the genus *Chonopeltis* Thiele, 1900 (Crustacea: Branchiura) with notes on its geographical distribution. Koedoe, **37**(1): 41-56.

AVENANT-OLDEWAGE, A. & E. KNIGHT, 2008. Aspects of the ecology of *Chonopeltis australis* Boxshall, 1976 in Boskop Dam, North West Province. South African Journal of Wildlife Research, **38**(1): 28-34.

AVENANT-OLDEWAGE, A. & W. H. OLDEWAGE, 1995. A new species of *Argulus* (Crustacea: Branchiura) from a bony fish in Algoa Bay, South Africa. South African Journal of Zoology, **30**(4): 197-199.

AVENANT-OLDEWAGE, A. & J. H. SWANEPOEL, 1993. The male reproductive system and mechanism of sperm transfer in *Argulus japonicus* (Crustacea: Branchiura). Journal of Morphology, **215**: 51-63.

AVENANT-OLDEWAGE, A., J. H. SWANEPOEL & E. KNIGHT, 1994. Histomorphology of the digestive tract of *Chonopeltis australis* (Crustacea: Branchiura). South African Journal of Zoology, **29**(1): 74-81.

AVENANT-OLDEWAGE, A. & J. G. VAN AS, 1990a. The digestive system of the fish ectoparasite *Dolops ranarum* (Crustacea: Branchiura). Journal of Morphology, **204**: 103-112.

AVENANT-OLDEWAGE, A. & J. G. VAN AS, 1990b. On the reproductive system of *Dolops ranarum* (Stuhlmann, 1891) (Crustacea: Banchiura). South African Journal of Zoology, **25**(1): 67-71.

BAIRD, W., 1850. The natural history of the British Entomostraca: 1-450. (The Ray Society, London).

BANDILLA, M., 2007. Transmission and host and mate location in the fish louse *Argulus coregoni* and its link with bacterial disease in fish. Jyväskylä studies in biological and environmental science, **179**: 1-40.

BANDILLA, M., T. HAKALAHTI, P. J. HUDSON & E. T. VALTONEN, 2005. Aggregation of *Argulus coregoni* (Crustacea: Branchiura) on rainbow trout (*Oncorhynchus mykiss*): a consequence of host susceptibility or exposure? Parasitology, **130**: 1-8.

BANDILLA, M., T. HAKALAHTI-SIRÉN & E. T. VALTONEN, 2007. Experimental evidence for a hierarchy of mate- and host-induced cues in a fish ectoparasite, *Argulus coregoni* (Crustacea: Branchiura). International Journal for Parasitology, **37**: 1343-1349.

BANDILLA, M., T. HAKALAHTI-SIRÉN & E. T. VALTONEN, 2008. Patterns of host switching in the fish ectoparasite *Argulus coregoni*. Behavioural Ecology and Sociobiology, **62**: 975-982.

BANDILLA, M., E. T. VALTONEN, L.-R. SUOMALAINEN, P. J. APHALO & T. HAKALAHTI, 2006. A link between ectoparasite infection and susceptibility to bacterial disease in rainbow trout. International Journal for Parasitology, **36**: 987-991.

BANERJEE, A., S. MANNA & S. K. SAHA, 2014a. Morphological characterization of testicular cells, spermatogenesis and formation of spermatophores in a fish ectoparasite *Argulus bengalensis* Ramakrishna, 1951 (Crustacea: Branchiura). Tissue Cell, **46**(1): 59-69.

BANERJEE, A., S. MANNA & S. K. SAHA, 2014b. Effect of aqueous extract of *Azadirachta indica* A. Juss (neem) leaf on oocyte maturation, oviposition, reproductive potentials and embryonic development of a freshwater fish ectoparasite *Argulus bengalensis* Ramakrishna, 1951 (Crustacea: Branchiura). Parasitology Research, **113**(12): 4641-4650.

BANERJEE, A., S. MANNA & S. K. SAHA, 2015. A histological evaluation of development and axis formation in freshwater fish ectoparasite *Argulus bengalensis* Ramakrishna, 1951 (Crustacea: Branchiura). Parasitology Research, **114**(6): 2199-2212.

BANERJEE, A. & S. K. SAHA, 2013. Tissue specific structural variations of mitochondria of fish ectoparasite *Argulus bengalensis* Ramakrishna, 1951 (Crustacea: Branchiura): functional implications. Journal of Advanced Research, **5**(3): 319-328.

BANERJEE, A. & S. K. SAHA, 2016. Histological and ultrastructural investigation of the female reproductive system of *Argulus bengalensis* Ramakrishna, 1951 (Crustacea: Branchiura). Journal of Morphology, **277**: 707-716.

BANERJEE, S. & P. K. BANDYOPADHYAY, 2010. Observation on prevalence of ectoparasites in carp fingerlings in two district of west Bengal. Journal of Parasitic Diseases, **34**(1): 44-47.

BANGHAM, R. V., 1940. Parasites of freshwater fish of southern Florida. Proceedings of the Florida Academy of Sciences, **5**: 289-307.

BANGHAM, R. V., 1941. Parasites from fish of Buckeye Lake, Ohio. Ohio Journal of Science, **41**(6): 441-448.

BANGHAM, R. V., 1946. Parasites of northern Wisconsin fish. Transactions of the Wisconsin Academy of Sciences, Arts, and Letters, **36**: 291-325.

BANGHAM, R. V. & C. E. VENARD, 1942. Studies on parasites of Reelfoot Lake fish. IV. Distribution studies and checklist of parasites. Journal of the Tennessee Academy of Science, **17**: 22-38.

BARNARD, K. H., 1955. South African parasitic Copepoda. Annals of the South African Museum, **41**: 228-312.

BAUER, O. M., 1962. Parasites of freshwater fish and the biological basis for their control. XLIX Translated from Russian. Israel Program for Scientific Translation, **49**: 642-673.

BAZAL, K., Z. LUCKY & V. DYK, 1969. Localization of fish-lice and leeches on carps during the autumn fishing. Acta Veterinaria (Brno), **38**: 533-544.

BENDA, R. S., 1969. Occurrence of *Argulus appendiculosus* Wilson, 1907 (Crustacea: Branchiura) in Indiana. Proceedings of the Indiana Academy of Science, **89**: 404.

BENDA, R. S., 1974. Occurrence of *Argulus mississippiensis* (Crustacea: Branchiura) in Indiana. Proceedings of the Indiana Academy of Science, **84**: 213-214.

BENZ, G. W., S. A. BULLARD & A. D. DOVE, 2001. Metazoan parasites of fishes: Synoptic information and portal to the literature for aquarists. In: Regional Conference Proceedings 2001, American Zoo and Aquarium Association, Silver Spring, MD.

BENZ, G. W., R. L. OTTING & A. CASE, 1995. Redescription of *Argulus melanostictus* (Branchiura: Argulidae), a parasite of California grunion (*Leuresthes tenuis*: Atherinidae), with notes regarding chemical control of *A. melanostictus* in a captive host population. The Journal of Parasitology, **81**(5): 754-761.

BERE, R., 1931. Copepods parasitic on fish of the Trout Lake region, with descriptions of two new species. Wisconsin Academy of Sciences, Arts, and Letters, **26**: 427-436.

BERE, R., 1936. Parasitic copepods from Gulf of Mexico fish. The American Midland Naturalist, **17**(3): 577-629.

BOOMKER, J. D. F., 1981. Some arthropods parasitic on freshwater fish. South African Journal of Science, **77**: 571.

BOUCHET, G. C., 1985. Redescription of *Argulus varians* Bere, 1936 (Branchiura, Argulidae) including a description of its early development and first larval stage. Crustaceana, **49**(1): 31-35.

BOUVIER, E.-L., 1897. Observations sur les Argulidés du genre *Gyropeltis* recueillis par M. Geay au Vénézuela. Bulletin du Museum national d'histoire naturelle, **3**: 13-19. [In French.]

BOUVIER, E.-L., 1899a. Les Crustacés parasites du genre *Dolops* Audouin (premiere Partie). Bulletin de la Societe Philomathique de Paris, **8**(10)2-3: 53-81. [In French.]

BOUVIER, E.-L., 1899b. Les Crustacés parasites du genre *Dolops* Audouin (Seconde Partie). Bulletin de la Societe Philomathique de Paris, **9**(1)1: 12-40. [In French.]

BOUVIER, E.-L., 1899c. Sur les Argulides du genre *Gyropeltis*, recueillis recemment par M. Geay dans la Guyane. Bulletin du Museum d'Histoire Naturelle, **5**(1): 39-41. [In French.]

BOUVIER, E.-L., 1910. Un Argulide nouveau de l'Argentine. *Argulus ichesi* nov. sp. Bulletin du Museum national d'histoire naturelle, 1ère Série, **16**(2): 92-95. [In French.]

BOWER-SHORE, C., 1940. An investigation of the common fish louse *Argulus foliaceus* (Linn.). Parasitology, **32**: 361-371.

BOXSHALL, G. A., 1976. A new species of *Chonopeltis* (Crustacea: Branchiura) from southern Africa. Bulletin of the British Museum (Natural History). Zoology, **30**: 217-221.

BOXSHALL, G. A., 2007. Crustacean classification: on-going controversies and unresolved problems. In: Z.-Q. ZHANG & W. A. SHEAR (eds.), Linnaeus tercentenary: progress in invertebrate taxonomy. Zootaxa, **1668**: 1-766.

BOXSHALL, G. A., 2009. *Argulus* O. F. Müller, 1785. In: T. C. WALTER & G. BOXSHALL (2016). World of Copepods database. World Register of Marine Species, available online at http://www.marinespecies.org/aphia.php?p=taxdetails&id=347072 (accessed 28 June 2016).

BOXSHALL, G. & T. C. WALTER, 2009. *Argulus vittatus* (Rafinesque-Schmaltz, 1814). In: T. C. WALTER & G. BOXSHALL (2016). World of Copepods database. World Register of Marine Species, available online at http://www.marinespecies.org/aphia.php?p=taxdetails&id=347072 (accessed 28 June 2016).

BOXSHALL, G. & T. C. WALTER, 2015. *Dipteropeltis* Calman, 1912. In: T. C. WALTER & G. BOXSHALL (2015). World of Copepods database. World Register of Marine Species, available online at http://www.marinespecies.org/aphia.php?p=taxdetails&id=347072 (accessed 28 June 2016).

BRIAN, A., 1924. Parasitologia Mauritanica, Matériaux pour la faune parasitologique en Mauritanie. Arthropoda (1re partie), Copepoda. Copépodes commensaux et parasites des côtes Mauritaniennes. Bulletin du Comité d'Études Historique et Scientifiques de l'Afrique Occidentale Française, **1924**: 1-66. [In French.]

BRIAN, A., 1927. Crustacea II. Copepod parasitica. Faune des Colonies Françaises, **1**(6): 572-587. [In French.]

BRIAN, A., 1940. Sur quelques Argulides d'Afrique appartenant aux collections du Musée du Congo Belge. Revue de Zoologie et de Botanique Africaines, **33**: 77-98. [In French.]

BRIAN, A., 1947. Los Argulidos del Museo Argentino de Ciencias Naturales (Crustacea Branchiura). Carcinologia, **15**: 353-370. [In French.]

BUCHMANN, K. & J. BRESCIANI, 1997. Parasitic infections in pond-reared rainbow trout *Oncorhynchus mykiss* in Denmark. Diseases of Aquatic Organisms, **28**: 125-138.

BUCHMANN, K., A. ULDAL & H. C. K. LYHOLT, 1995. A checklist of metazoan parasites from rainbow trout (*Oncorhynchus mykiss*). Acta Veterinaria Scandinavica, **36**(3): 299-318.

BUNKLEY-WILLIAMS, L. & E. H. WILLIAMS, 1994. Parasites of Puerto Rican freshwater sport fishes. (Puerto Rico Department of Natural and Environmental Resources, San Juan, PR and Department of Marine Sciences, University of Puerto Rico, Mayaguez, PR).

BUTTNER, J. K., 1980. Occurrence of *Argulus* (Crustacea: Branchiura) on the white crappie, *Pomoxis annularis* Rafinesque, and free-swimming in the southern Illinois reservoirs. Transactions of the Illinois State Academy of Science, **72**(3): 5-7.

BYRNES, T., 1985. Two new *Argulus* species (Branchiura: Argulidae) found on Australian bream (*Acanthopagrus* spp.). Australian Journal of Zoology, **21**(7): 579-586.

CALMAN, W. T., 1912. On *Dipteropeltis*, a new genus of the crustacean order Branchiura. Proceedings of the Zoological Society of London, 763-766.

CAMPBELL, A. D., 1971. The occurrence of *Argulus* (Crustacea: Branchiura) in Scotland. Journal of Fish Biology, **3**: 145-146.

CANIĆ, J., V. STOJADINOVIĆ & N. ZIVKOVIĆ, 1977. *Argulus foliaceus* (Linné, 1758) the causal agent of deaths of the carp offsprings in a fishpond. Veterinarski glasnik Belgrade, **4**: 297-298.

CARVALHO, L. N., R. ARRUDA & K. DEL-CLARO, 2004. Host-parsite interactions between the piranha *Pygocentrus nattereri* (Characiformes: Characidae) and isopods and branchiurans (Crustacea) in the rio Araguaia basin, Brazil. Neotropical Ichthyology, **2**(2): 93-98.

CARVALHO, L. N., K. DEL-CLARO & R. M. TAKEMOTO, 2003. Host-parasite interaction between branchiurans (Crustacea: Argulidae) and piranhas (Osteichthyes: Serrasalminae) in the Pantanal wetland of Brazil. Environmental Biology of Fishes, **67**: 289-296.

CAUSEY, D., 1960. Parasitic Copepoda from Mexican coastal fishes. Bulletin of Marine Science of the Gulf and Caribbean, **10**: 323-337.

CENGIZLER, I., N. AYTAC, A. SAHAN, A. A. OZAK & E. GENÇ, 2001. Ecto-endo parasite investigation on mirror carp (*Cyprinus carpio* L., 1758) captured from the River Seyhan, Turkey. E.U. Journal of fisheries and Aquatic Sciences, **18**(1-2): 87-90.

CHEN, T.-P., 1933. A study of the methods of prevention and treatment of fish lice in pond culture. Lingnan Science Journal, **12**(2): 241-244.

CHOWDHURY, M. M., M. D. RAKNUZZAMAN & K. F. IQUBAL, 2006. Control of *Argulus* sp. infestation in goldfish (*Carrassius auratus*) with Sumithion. Bangladesh Journal of Zoology, **34**(1): 111-115.

CLAUS, C., 1875. Ueber die Entwicklung, Organisation und systematische Stellung der Arguliden. Zeitschrift für Wissenschaftliche Zoologie, **25**: 217-284. [In German.]

CORNALIA, E., 1860. Sopra una nuova specie di crostacei sifonostomi. Memorie del Rey Istituto lombardo di scienze e lettere, **3**: 161-171. [In Italian.]

CRESSEY, R. F., 1971. Two new argulids (Crustacea: Branchiura) from the eastern United States. Proceedings of the Biological Society of Washington, **84**(31): 253-258.

CRESSEY, R. F., 1972. The genus *Argulus* (Crustacea: Branchiura) of the United States. Biota of freshwater ecosystems, Identification Manual 2.

CRESSEY, R. F., 1978. Marine Flora and Fauna of Northeastern United States. Crustacea: Branchiura. NOAA Technical Report Circular 413.

CUÉNOT, L., 1912. Contributions a la faune du Bassin d'Arcachon. VI. Argulides. Description d'*Argulus arcassonensis*, nov. sp. Bulletin de la Station de Biologie d'Arcachon, **14**: 117-125. [In French.]

CUNNINGTON, W. A., 1913. Zoological results of the Third Tanganyika Expedition conducted by Dr W. A. Cunnington, 1904-1905. Report on the Branchiura. Proceedings of the Zoological Society of London, **18**: 262-283.

CUNNINGTON, W. A., 1931. Reports of an expedition to Brazil and Paraguay in 1926-1927, supported by the Trustees of the Percy Sladen Memorial Fund and the Executive Committee of the Carnegie Trust for Scotland. Argulidae. The Journal of the Linnean Society of London. Zoology, **37**: 259-263.

CZECZUGA, B., 1971. Comparative studies on the occurrence of carotenoids in the host-parasite system, with reference to *Argulus foliaceus* (L.) (Crustacea) and its host *Gasterosteus aculeatus* L. (Pisces). Acta Parasitologica Polonica, Warsaw, **19**(15): 185-194.

DA SILVA, N. M. M., 1978. Uma nova espécie de crustáceo argulideo no Rio Grande do Sul, Brasil (Branchiura, Argulidae). Iheringia Serie Zoologia, Porto Alegre (52): 3-29. [In Portuguese.]

DA SILVA, N. M. M., 1980. *Argulus spinulosus* sp. n. (Branchiura, Argulidae), em peixes de água doce do Rio Grande do Sul, Brasil. Iheringia Serie Zoologia, Porto Alegre (56): 15-23. [In Portuguese.]

DANA, J. D., 1853. Crustacea. Part 2. In: United States Exploring Expedition during the years 1838, 1839, 1840, 1841, 1842 under the command of Charles Wilkes, U.S.N., **14**: 389-1618.

DANA, J. D. & E. C. HERRICK, 1837. Description of the *Argulus catastomi*, a new parasitic crustaceous animal (with figures). The American Journal of Science and Arts, **31**: 297-308.

DARTEVELLE, E., 1951. Crustacés de poisons du Congo. Zooleo (N. S.), **9**: 11-13. [In French.]

DAVIS, C. C., 1965. A study of the hatching process in aquatic invertebrates. XVII. Hatching in *Argulus megalops* Smith (Crustacea: Branchiura). Hydrobiologia, **27**(1-2): 202-207.

DEBAISIEUX, P., 1953. Histologie et histogenèse chez *Argulus foliaceus* L. (Crustacé, Branchiure). Cellule, **55**: 1-53. [In French.]

DECHTIAR, A. O., 1972. New parasite records for Lake Erie fish. Great Lakes Fishery Commission Technical Report, **17**: 1-20.

DECHTIAR, A. O. & W. J. CHRISTIE, 1988. Survey of the parasite fauna of Lake Ontario fishes, 1961-1971. In: S. J. NEPZY (ed.), Parasites of fishes in the Canadian waters of the Great Lakes: 66-95. Great Lakes Fishery Commission Technical Report 51.

DECHTIAR, A. O., J. J. COLLINS & J. A. RECKAHN, 1988. Survey of the parasite fauna of Lake Huron Fishes, 1961 to 1971. In: S. J. NEPZY (ed.), Parasites of fishes in the Canadian waters of the Great Lakes: 19-48. Great Lakes Fishery Commission Technical Report 51.

DECHTIAR, A. O. & S. J. NEPZY, 1988. Survey of the parasite fauna of selected fish species from Lake Erie, 1970-1975. In: S. J. NEPZY (ed.), Parasites of fishes in the Canadian waters of the Great Lakes: 49-65. Great Lakes Fishery Commission Technical Report 51.

DEUTSCH, W. G., 1989. Parasites of Susquehanna River (Pennsylvania) Muskellunge. Journal of the Pennsylvania Academy of Science, **63**(1): 25-27.

DEVERAJ, M. & K. M. S. AMEER HAMSA, 1977. A new species of *Argulus* (Branchiura) from a marine fish, *Psammoperca waigiensis* (Cuvier). Crustaceana, **32**(2): 129-134.

DOLLFUS, R. P. H., 1960. VII. Copepodes parasites de Teleosteens du Niger. Bulletin de l'Institut Fondamental d'Afrique Noire, Series A (1): 170-192. [In French.]

DOUËLLOU, L. & K. H. ERLWANGER, 1994. Crustacean parasites of fishes in Lake Kariba, Zimbabwe, preliminary results. Hydrobiologia, **287**: 233-242.

DUGATKIN, L. A., G. J. FITZGERALD & J. LAVOIE, 1994. Juvenile three-spined sticklebacks avoid parasitized conspecifics. Environmental Biology of Fishes, **39**: 215-218.

DUTCHER, B. W. & F. J. SCHWARTZ, 1962. A preferential parasitic copepod-oyster Toadfish Association. Chesapeake Science, **3**(3): 213-215.

EDGECOMB, G. D., 2010. Arthropod phylogeny: an overview from the perspectives of morphology, molecular data and the fossil record. Arthrop. Structure and Development, **39**: 74-87.

EVERTS, L. & A. AVENANT-OLDEWAGE, 2009. First record of *Argulus coregoni*: a fish ectoparasitic crustacean from Malaysia and additional notes on the morphology. Malaysian Applied Biology, **38**(2): 61-71.

FENTON, A., T. HAKALAHTI, M. BANDILLA & E. T. VALTONEN, 2006. The impact of variable hatching rates on parasite control: a model of an aquatic ectoparasite in a Finnish fish farm. Journal of Applied Ecology, **43**: 660-668.

FISCHTHAL, J. H., 1947. Parasites of northwest Wisconsin fishes I. The 1944 survey. Transactions of the Wisconsin Academy of Sciences, Arts, and Letters, **37**: 157-220.

FISCHTHAL, J. H., 1950. Parasites of northwest Wisconsin fishes II. The 1945 survey. Transactions of the Wisconsin Academy of Sciences, Arts, and Letters, **40**: 87-113.

FISCHTHAL, J. H., 1952. Parasites of northwest Wisconsin fishes III. The 1946 survey. Transactions of the Wisconsin Academy of Sciences, Arts, and Letters, **41**: 17-58.

FONTANA, M., R. M. TAKEMOTO, J. C. O. MALTA & L. A. F. MATEUS, 2012. Parasitism by argulids (Crustacea: Branchiura) in piranhas (Osteichthyes: Serrasalmidae) captured in the Caiçara bays, upper Paraguay River, Pantanal, Mato Grosso State, Brazil. Neotropical Ichthyology, **10**(3): 653-659.

FORLENZA, M., P. D. WALKER, B. J. DE VRIES, S. E. WENDELAAR BONGA & G. F. WIEGERTJES, 2008. Transcriptional analysis of the common carp (*Cyprinus carpio* L.) immune response to the fish louse *Argulus japonicus* Thiele (Crustacea: Branchiura). Fish and Shellfish Immunology, **25**: 76-83.

FOX, H. M., 1957. Hæmoglobin in Branchiura. Nature, **179**: 873.

FRYER, G., 1956. A report on the parasitic Copepoda and Branchiura of the fishes of Lake Nyasa. Proceedings of the Zoological Society of London, **127**: 293-344.

FRYER, G., 1958. Occurrence of spermatophores in the genus *Dolops* (Crustacea: Branchiura). Nature, **181**: 1011-1012.

FRYER, G., 1959. A report on the parasitic Copepoda and Branchiura of the fishes of Lake Bangweulu (Northern Rhodesia). Proceedings of the Zoological Society of London, **132**: 517-550.

FRYER, G., 1960a. Studies on some parasitic crustaceans on African freshwater fishes, with descriptions of a new copepod of the genus *Ergasilus* and a new branchiuran of the genus *Chonopeltis*. Proceedings of the Zoological Society of London, **133**(4): 629-647.

FRYER, G., 1960b. The spermatophores of *Dolops ranarum* (Crustacea: Branchiura): their structure, formation, and transfer. Quarterly Journal of Microscopical Science, **101**(4): 407-432.

FRYER, G., 1961a. The parasitic Copepoda and Branchiura of the fishes of Lake Victoria and the Victoria Nile. Proceedings of the Zoological Society of London, **137**(1): 41-60.

FRYER, G., 1961b. Larval development in the genus *Chonopeltis* (Crustacea: Branchiura). Proceedings of the Zoological Society of London, **137**: 61-69.

FRYER, G., 1963. Crusatacean parasites from cichlid fishes of the genus *Tilapia* in the Musée Royal de l'Afrique central. Revue de Zoologie et de Botanique Africaines, **68**(3-4): 386-392.

FRYER, G., 1964. Further studies on the parasitic Crustacea of African freshwater fishes. Proceedings of the Zoological Society of London, **143**: 79-102.

FRYER, G., 1965a. Crustacean parasites of African freshwater fishes mostly collected during the expeditions to Lake Tanganyika, and to Lakes Kivu, Edward and Albert by the Institut Royal des Sciences Naturelles de Belgique. Bulletin Institut Royal des Sciences naturelles de Belgique, **XLI**(7): 1-22.

FRYER, G., 1965b. Parasitic crustaceans of African freshwater fishes from the Nile and Niger systems. Proceedings of the Zoological Society of London, **145**: 285-303.

FRYER, G., 1968. The parasitic Crustacea of African freshwater fishes; their biology and distribution. Journal of Zoology London, **156**: 45-95.

FRYER, G., 1969. A new freshwater species of the genus *Dolops* (Crustacea: Branchiura) Parasitic on a galaxiid fish of Tasmania — with comments on disjunct distribution patterns in the Southern Hemisphere. Australian Journal of Zoology, **17**: 49-64.

FRYER, G., 1974. Une nouvelle espèce de *Chonopeltis* (Crustacea: Branchiura), parasite d'un poisson congolais. Revue de Zoologie Africaine, **88**(2): 437-440.

FRYER, G., 1977. On some species of *Chonopeltis* (Crustacea: Branchiura) from the rivers of the extreme South West Cape region of Africa. Journal of Zoology London, **182**(1): 441-455.

FRYER, G., 1978. Parasitic Copepoda and Branchiura. In: J. ILLIES (ed.), Limnofauna Europaea. A checklist of the animals inhabiting European inland waters, with account of their distribution and ecology (2nd revised and enlarged ed.). (G. Fischer, Stuttgart and Swets & Zeitlinger, Amsterdam).

FRYER, G., 1982. The parasitic Copepoda and Branchiura of British freshwater fishes. Freshwater Biological Association Scientific Publication, **46**. (Freshwater Biological Association, Windmere).

GATTAPONI, P., 1971. *Argulus foliaceus* L. (Crustacea: Argulidae) su pesci del bacino del Lago Trasimeno. Atti della Societa Italiana delle Scienze Veterinarie, **25**: 471-473. [In Italian.]

GAULT, N. F. S., D. J. KILPATRICK & M. T. STEWART, 2002. Biological control of the fish louse in a rainbow trout fishery. Journal of Fish Biology, **60**: 226-237.

GOIN, C. J. & L. H. OGREN, 1956. Parasitic copepods (Argulidae) on amphibians. The Journal of Parasitology, **42**(2): 172.

GOMES, A. L. S. & J. C. O. MALTA, 2002. Postura, desenvolvimento e eclosão dos ovos de *Dolops carvalhoi* Lemos de Castro (Crustácea, Branchiura) em laboratório, parasite de peixes da Amazônia Central. Revista Brasileira de Zoologia, **19**(Supl. 2): 141-149. [In Portuguese.]

GOULD, A. A., 1841. A report on the Invertebrata of Massachusetts, comprising the Mollusca, Crustacea, Annelida, and Radiata. (Folsom, Wells and Thurston, Cambridge).

GRABDA, J., 1971. Catalogue of Polish parasiite fauna. Part II. Parasites of Cyclostoma and fish: 9-304. (PWN, Warsaw). [In Polish.]

GRESTY, K. A., G. A. BOXSHALL & K. NAGASAWA, 1993. The fine structure and function of the cephalic appendages of the branchiuran parasite, *Argulus japonicus* Thiele. Philosophical Transactions of the Royal Society Series B: Biological Sciences, **339**: 119-135.

GROBBEN, K., 1908. Beiträge zur Kenntnis des Baues und der systematischen Stellung der Arguliden. Sitzungsberichte der Kaiserlichen Akademie der Wissenschaften in Wien, **117**: 191-233. [In German.]

GUBERLET, J. E., 1928. Notes on a species of *Argulus* from gold-fish. Publications in fisheries, University of Washington, College of fisheries, **2**(3): 31-42.

GUHA, A., G. ADITYA & S. K. SAHA, 2012. Correlation between body size and fecundity in fish louse *Argulus bengalensis* Ramakrishna, 1951 (Crustacea: Branchiura). Journal of Parasitic Diseases, **37**(1): 118-124.

GUHA, A., G. ADITYA & S. K. SAHA, 2013. Survivorship and fecundity of *Argulus bengalensis* (Crustacea; Branchiura) under laboratory conditions. Invertebrate Reproduction & Development, **57**(4): 301-308.

GUIDELLI, G., W. L. G. TAVECHIO, R. M. TAKEMOTO & G. C. PAVANELLI, 2006. Parasite fauna of *Leporinus lacustris* and *Leporinus friderici* (Characiformes, Anostomidae) from the upper Paraná River floodplain, Brazil. Revista Tecnologica (Maringá, Brazil), **28**(3): 281-290. [In Portuguese with English summary.]

GURNEY, R., 1948. The British species of fish-louse of the genus *Argulus*. Proceedings of the Zoological Society of London, **118**(3): 553-558.

HAASE, W., 1974. Ein bisher nicht erkanntes Chloridzellen-Organ der Karpfenlaus *Argulus foliaceus* L. (Crustacea, Branchiura). Experientia, **30**(4): 407-409. [In German.]

HAASE, W., 1975a. Ultrastruktur und Funktion der Carapaxfelder von *Argulus foliaceus* L. Zeitschrift für Morphologie der Tiere, **81**: 161-189. [In German.]

HAASE, W., 1975b. Ultrastrukturelle Veränderungen der Carapaxfelder von *Argulus foliaceus* L. in Abhängigkeit vom Ionengehalt des Lebensraums (Crustacea, Branchiura). Zeitschrift für Morphologie der Tiere, **81**: 343-353. [In German.]

HAECKEL, E., 1866. Generelle Morphologie der Organismen: allgemeine Grundzuge der organischen Formen-Wissenschaft, mechanisch begründet durch die von Charles Darwin reformirte Descendenz-Theorie: 1-626. (Georg Reimer, Berlin). [In German.]

HAKALAHTI, T., M. BANDILLA & E. T. VALTONEN, 2005. Delayed transmission of a parasite is compensated by accelerated growth. Parasitology, **131**: 647-656.

HAKALAHTI, T., H. HÄKKINEN & E. T. VALTONEN, 2004a. Ectoparasitic *Argulus coregoni* (Crustacea: Branchiura) hedge their bets — studies on egg hatching dynamics. Oikos, **107**: 295-302.

HAKALAHTI, T., Y. LANKINEN & E. T. VALTONEN, 2004b. Efficacy of emamectin benzoate in the control of *Argulus coregoni* (Crustacea: Branchiura) on rainbow trout *Oncorhynchus mykiss*. Diseases of Aquatic Organisms, **60**: 197-204.

HAKALAHTI, T., A. F. PASTERNAK & E. T. VALTONEN, 2004c. Seasonal dynamics of egg laying and egg-laying strategy of the ectoparasite *Argulus coregoni* (Crustacea: Branchiura). Parasitology, **128**: 655-660.

HAKALAHTI, T. & E. T. VALTONEN, 2003. Population structure and recruitment of the ectoparasite *Argulus coregoni* Thorell (Crustacea: Branchiura) on a fish farm. Parasitology, **127**: 79-85.

HALLBERG, E., 1982. The fine structure of the compound eye of *Argulus foliaceus* (Crustacea: Branchiura). Zoologischer Anzeiger Jena, **208**(3-4): 227-236.

HAN, K.-S., Y.-M. JUNG, T.-W. PARK, C.-W. LIM, H.-J. SONG & H.-K. DO, 1998. Experimental infection of *Argulus japonicus* in freshwater fishes. Korean Journal of Veterinary Service, **21**(4): 431-437.

HANSON, S. K., J. E. HILL, C. A. WATSON, R. P. E. YANONG & R. ENDRIS, 2011. Evaluation of Emamectin Benzoate for the control of experimentally induced infestations of *Argulus* sp. in goldfish and koi carp. Journal of Aquatic Animal Health, **23**(1): 30-34.

HAOND, C., D. T. NOLAN, N. M. RUANE, J. ROTLLANT & S. E. WENDELAAR BONGA, 2003. Cortisol influences the host-parasite interaction between the rainbow trout (*Oncorhynchus mykiss*) and the crustacean ectoparasite *Argulus japonicus*. Parasitology, **127**: 551-560.

HARGIS, W. J., 1958. The fish parasite *Argulus laticauda* as a fortuitous human epizoon. The Journal of Parasitology, **44**(1): 45.

HARRISON, A. J., N. F. S. GAULT & J. T. A. DICK, 2006. Seasonal and vertical patterns of egg-laying by the freshwater fish louse *Argulus foliaceus* (Crustacea: Branchiura). Diseases of Aquatic Organisms, **68**: 167-173.

HARRISON, A. J., N. F. S. GAULT & J. T. A. DICK, 2007. Diel variation in egg-laying by the freshwater fish louse *Argulus foliaceus* (Crustacea: Branchiura). Diseases of Aquatic Organisms, **78**: 169-172.

HEEGARD, P., 1962. Parasitic Copepoda from Australian waters. Records of the Australian Museum, **25**(9): 149-233.

HELLER, C., 1857. Beiträge zur Kenntniss der Siphonostomen. Sitzungsberichte der Akademie der Wissenschaften mathematisch-naturwissenschaftliche Klasse, **25**: 89-108. [In German.]

HEMAPRASANTH, K. P., B. KAR, S. K. GARNAYAK, J. MOHANTY, J. K. JENA & P. K. SAHOO, 2012. Efficacy of two avermectins, doramectin and ivermectin against *Argulus siamensis* infestation in Indian major carp, *Labeo rohita*. Veterinary Parasitology, **190**: 297-304.

HERTER, K., 1926. Reizphysiologische Untersuchungen an der Karpfenlaus (*Argulus foliaceus* L.). Zeitschrift für vergleichende Physiologie, **5**(2): 283-370. [In German.]

HINDLE, E., 1948. Notes on the treatment of fish infected with *Argulus*. Proceedings of the Zoological Society of London, **119**: 79-81.

HINES, R., 1972. Some common diseases of carp in Israel. Refuah Veterinarith, **29**(1): 37-44.

HOFFMAN, G. L., 1967. Parasites of North American freshwater fishes. (University of California Press, Berkeley, CA).

HOFFMAN, G. L., 1977. *Argulus*, a branchiuran parasite of freshwater fishes. US Fish and Wildlife Publications, **137**: 2-9.

HOLTHUIS, L. B., 1954. C. S. Rafinesque as a carcinologist, an annotated compilation of the information on Crustacea contained in the works of that author. Zoologische Verhandelingen, Leiden, **25**: 1-43.

HOSHINA, T., 1950. Über eine *Argulus*-art im Salmonidenteiche. Bulletin of the Japanese Society of Scientific Fisheries, **16**: 239-243. [In German with English summary.]

HSIAO, S. C., 1950. Copepods from Lake Erh Hai, China. Proceedings of the United States National Museum, **100**(3261): 161-200.

HUGGHINS, E. J., 1970. Argulids (Crustacea: Branchiura) from Ecuador and Bolivia. The Journal of Parasitology, **56**(5): 1003.

ICZN, 1954. Opinion 288. Designation, under the plenary powers, of the generic name *Monoculus* Linnaeus, 1758 (systematic position indeterminate) are surpressed by the laws of priority but not for those of the law of homonymy. Opinions and Declarations rendered by the International Commission on Zoological Nomenclature, **8**(5): 63-72.

IDER, D., Z. RAMDANE, L. COURCOT, R. AMARA & J.-P. TRILLES, 2014. A scanning electron microscopy study of *Argulus vittatus* (Rafinesque-Schmaltz, 1814) (Crustacea: Branchiura) from Algerian coast. Parasitology Research, **113**(6): 2265-2276.

IKUTA, K. & T. MAKIOKA, 1992. Structure of the female reproductive system in *Argulus japonicus* (Crustacea: Branchiura). Zoological Science, **9**(6): 1283.

IKUTA, K. & T. MAKIOKA, 1994. Notes on the postembryonic development of the ovary in *Argulus japonicus* (Crustacea: Branchiura). Proceedings of the Arthropodan Embryological Society of Japan, **29**: 15-17.

IKUTA, K. & T. MAKIOKA, 1997. Structure of the adult ovary and oogenesis in *Argulus japonicus* Thiele (Crustacea: Branchiura). Journal of Morphology, **231**: 29-39.

IKUTA, K., T. MAKIOKA & R. AMIKURA, 1997. Eggshell ultrastructure in *Argulus japonicus* (Branchiura). Journal of Crustacean Biology, **17**(1): 45-51.

INOUE, K., S. SHIMURA, M. SAITO & K. NISHIMURA, 1980. The use of trichlorfon in the control of *Argulus coregoni*. Fish Pathology, **15**(1): 37-42. [In Japanese with English summary.]

JAFRI, S. I. H. & S. S. AHMED, 1991. A new record of ectoparasitic crustaceans (Branchiura: Argulidae) from major carps in Sindh, Pakistan. Pakistan Journal of Zoology, **23**(1): 11-13.

JALALI, B. & M. BARZEGAR, 2006. Fish parasites in Zarivar Lake. Journal of Agricultural Science and Technology (Tehran, Islamic Republic of Iran), **8**: 47-58.

JALALI, B., M. BARZEGAR & H. NEZAMABADI, 2008. Parasitic fauna of the spiny eel, *Mastacembelus mastacembelus* Banks et Solander (Teleostei: Mastacembelidae) in Iran. Iranian Journal of Veterinary Research, Shiraz University, **9**(2/23): 158-161.

JÍROVEC, O. & K. WENIG, 1934. Das Atmungsorgan von *Argulus foliaceus* L. Zeitschrift für vergleichende Physiologie, **20**(4): 450-453. [In German.]

JURINE, M., 1806. Mémoire sur l'Argule foliacé (*Argulus foliaceus*). Annales du Muséum d'Histoire Naturelle, **7**: 431-548. [In French.]

KAJI, T., O. S. MØLLER & A. TSUKAGOSHI, 2011. A bridge between original and novel states: ontogeny and function of "suction discs" in the Branchiura (Crustacea). Evolution & Development, **13**(2): 119-126.

KARATOY, E. & E. SOYLU, 2006. Parasites of bream (*Abramis brama* Linnaeus, 1758) in the Lake Durusu (Terkos). Türkiye Parazitoloji Dergisi, **30**(3): 233-238. [In Turkish with English summary.]

KELLICOTT, D. S., 1877. Description of a new species of *Argulus*. Bulletin of the Buffalo Society of Natural Sciences, **3**(1875-1877): 214-216.

KELLICOTT, D. S., 1880. *Argulus stizostethii*, n.s. The American Journal of Microscopy, and Popular Science, **5**(3): 53-58.

KELLICOTT, D. S., 1886. A note on *Argulus catastomi*. Proceedings of the American Society of Microscopists, **8**: 144.

KENNEDY, C. R., 1975. The distribution of some crustacean fish parasites in Britain in relation to the introduction and movement of freshwater fish. Aquaculture Research, **6**(2): 36-41.

KHALIFA, K. A., 1989. Incidence of parasitic infestation of fishes in Iraq. Pakistan Veterinary Journal, **9**(2): 66-69.

KIRTISINGHE, P., 1959. A new marine *Argulus* (Copepoda, Branchiura). Ceylon Journal of Science, Biological Series, **2**(2): 253-255.

KIRTISINGHE, P., 1964. A review of the parasitic copepods of fish recorded from Ceylon with descriptions of additional forms. Bulletin of the Fisheries Research Station, Ceylon, **17**(1): 45-132.

KNIGHT, E. & A. AVENANT-OLDEWAGE, 1990. Aspects of the anatomy and morphology of the digestive tract of *Chonopeltis australis* (Crustacea: Branchiura). Electron microscopy society of southern Africa, **20**: 199-200.

KNUCKLES, J. L., 1972. *Ictalurus melas* (Rafinesque), a new host for *Argulus diversus* (Wilson). The Journal of Parasitology, **58**(3): 624.

KNUCKLES, J. L., 1977. *Lepomis macrochirus* Rafinesque, a new host for *Argulus maculosus* Wilson. The Journal of Parasitology, **63**(2): 383.

KNUCKLES, J. L., 1978. *Chaenobryttus coronaries* (Bartram), a new host for *Argulus maculosus* Wilson. Journal of Parasitology, **64**(5): 802.

KNUCKLES, J. L., 1980. *Argulus maculosus* (Wilson) from the pumpkinseed, *Lepomis gibbosus* (Linnaeus). The Journal of Parasitology, **66**(6): 919.

KNUCKLES, J. L., 1984. The shell cracker, *Lepomis microlophus* (Gunther), a new host for *Argulus maculosus* (Wilson). The Journal of Parasitology, **70**(3): 371.

KØIE, M., 1988. Parasites in eels, *Anguilla Anguilla* (L.), from eutrophic Lake Esrum (Denmark). Acta Parasitologica Polonica, **33**(2): 89-100.

KONDO, M., S. TOMONAGA & Y. TAKAHASHI, 2003. Morphological and cytochemical properties of hemocyte of *Argulus japonicus* (Arguloida, Branchiura, Crustacea). Journal of National Fisheries University, **51**(2): 45-52. [In Japanese with English summary.]

KOYUN, M., 2011. The effect of water temperature on *Argulus foliaceus* L. 1758 (Crustacea; Branchiura) on different fish species. Notulae Scientia Biologicae, **3**: 16-19.

KROGER, R. L. & J. F. GUTHRIE, 1972. Occurrence of the parasitic branchiuran, *Argulus alosae*, on dying Atlantic menhaden, *Brevoortia tyrannus*, in the Connecticut River. Transactions of the American Fisheries Society, **3**: 559-560.

KRØYER, H., 1863. Bidrag til Kunskab om Snyltekrebsene. Naturhistorisk Tidsskrift, **3**: 1-352. [In Danish.]

KRUGER, I., J. G. VAN AS & J. E. SAAYMAN, 1983. Observations on the occurrence of the fish louse *Argulus japonicus* Thiele, 1900 in the western Transvaal. South African Journal of Zoology, **18**(4): 408-410.

KU, C.-T. & K.-N. WANG, 1956. Studies on a new species of *Argulus* (Crustacea) with its larval stages found on the yellow-barbed catfish, in Tientsin, China. Tung Wu Hsüeh Pac, **8**(1): 41-47. [In Chinese with English summary.]

KUBRAKIEWICZ, J. & M. KLIMOWICZ, 1994. The ovary structure in *Argulus foliaceus* (Branchiura; Crustacea). Zoologica Poloniae, **39**(1-2): 7-14.

KUMAR, S., T. S. KUMAR, R. VIDYA & P. K. PANDEY, 2016. A prospective of epidemiological intervention in investigation and management of argulosis in aquaculture. Aquaculture International, in press. DOI:10.1007/s10499-016-0030-0.

KUMAR, S., R. P. RAMAN, K. KUMAR, P. K. PANDEY, N. KUMAR, B. MALLESH, S. MOHANTY & A. KUMAR, 2013. Effect of azadirachtin on haematological and biochemical parameters of *Argulus*-infested goldfish *Carassius auratus* (Linn. 1758). Fish Physiology and Biochemistry, **39**: 733-747.

KUMAR, A., R. P. RAMAN, K. KUMAR, P. K. PANDEY, V. KUMAR, S. MOHANTY & S. KUMAR, 2012. Antiparasitic efficacy of piperine against *Argulus* spp. on *Carassius auratus* (Linn. 1758): in vitro and in vivo study. Parasitology Research, **111**: 2071-2076.

LAHILLE, F., 1926. Nota sobre unos parasitos de los bagres dorados y surubies. Revista del Centro Estudiantes de Agronomia y Veterinaria, Buenos Aires, **19**: 6-16. [In Spanish.]

LAMARRE, E. & P. A. COCHRAN, 1992. Lack of host species selection by the exotic parasitic crustacean, *Argulus japonicus*. Journal of freshwater ecology, **7**(1): 77-80.

LANDSBERG, J. H., 1989. Parasites and associated diseases of fish in warm water culture with special emphasis on intensification. In: M. SHILO & S. SARIG (eds.), Fish culture in warm water systems: problems and trends: 195-252. (CRC Press, Boca Raton, FL).

LATREILLE, P. A., 1802. Histoire naturelle, générale et particulière des crustacés et des insectes: ouvrage faisant suite aux oeuvres de Leclerc de Buffon, et partie du cours complet d'histoire naturelle rédigé. (F. Dufart, Paris). An X-XIII, p. 914. [In French.]

LAURENT, P. J., 1975. Les argulues du Léman. Hydrologie, **37**(2): 249-252.

LEACH, W. E., 1819. Entomostracés, Entomostraca. Dictionnaire des Sciences Naturelles, **14**: 524-543. [In French.]

LEIGH-SHARPE, H., 1933. *Argulus rothschildi* n. sp., a parasitic Branchiura of *Abramis brama*. Parasitology, **24**: 552-557.

LEMOS DE CASTRO, A., 1949. Contribuição ao conhecimento dos crustaceos Argulideos do Brasil (Branchiura Argulidae), com descrição de uma nova espécie. Boletim do Museu Nacional Nova Serie Zoologia, **93**: 1-7. [In Portuguese.]

LEMOS DE CASTRO, A., 1950. Contribuição ao conhecimento dos Crustáceos Argulídeos do Brasil II. Descrição de duas novas espécies. An. da Acad. Brasileira de Ciências, **22**(2): 245-255. [In Portuguese.]

LEMOS DE CASTRO, A., 1951. Descrição do alótipo macho de *Argulus multicolour* Stekhoven, 1937 (Branchiura, Argulidae). Arquivos do Museu Nacional do Rio de Janeiro, **42**: 159-160. [In Portuguese.]

LEMOS DE CASTRO, A. & M. M. GOMES-CORRÈA, 1985. *Argulus hylae*, especie nova de Argulidae parasite de Girino. In: XII Congresso Brasileiro de Zoologia, Resumos. 27 de Janeiro a 1 de Fevereiro de 1985, Universidade Estadual de Campinas, São Paulo. p. 52. [In Portuguese.]

LESTER, R. J. G. & C. J. HAYWARD, 2006. Phylum Arthropoda. In: P. T. K. WOO (ed.), Fish diseases and disorders **1**: 466-565. (CAB international, Cambridge).

LEYDIG, F., 1850. Ueber *Argulus foliaceus*. Ein Beitrag zur Anatomie, Histologie und Entwicklungsgeschichte dieses Thieres. Zeitschrift für Wissenschaftliche Zoologie, **2**: 323-349. [In German.]

LEYDIG, F., 1871. Über einen *Argulus* der Umgebung von Tübingen. Archiv für Naturgeschichte, **37**(1): 1-23. [In German.]

LINNAEUS, C., 1758. Systema Naturae, per regna tria naturae, secundum Classes, Ordines, Genera, Species, cum characteribus, differentiis, synonymis, locis. Ed. decima, reformata. I, Regnum animale. Laurentius Salvius, Holmiae. [In Latin.]

LUCAS, P. H., 1849. Histoire naturelle des animaux articulés. Première partie: Crustacés, Arachnides, Myriapodes et Hexapodes. In: Exploration scientifique de l'Algérie pendant les années 1840, 1841, 1842; publiée par ordre du gouvernement et avec le concours d'une commission académique. Sciences physiques, zoologie. [In French.]

LUTSCH, E. & A. AVENANT-OLDEWAGE, 1995. The ultrastructure of the newly hatched *Argulus japonicus* Thiele, 1900 larvae (Branchiura). Crustaceana, **68**(3): 329-340.

LUUS-POWELL, W. J. & A. AVENANT-OLDEWAGE, 1996. Surface morphology of the fish parasite *Chonopeltis victori* Avenant-Oldewage, 1991 and aspects of the histomorphology. Koedoe, **39**(1): 55-70.

MACCHIONI, F., L. CHELUCCI, B. TORRACCA, M. C. PRATI & M. MAGI, 2015. Parasites of introduced goldfish (*Carassius auratus* L.) in the Massaciuccoli water district (Tuscany, Central Italy). Bulletin of the European Association of Fish Pathologists, **35**(2): 35-40.

MADSEN, N., 1964. The anatomy of *Argulus foliaceus* Linne. Lunds Universitets Arsskrift, **59**(13): 1-31.

MAHAR, M. A. & S. I. H. JAFRI, 2011. A new ectoparasitic crustacean, *Argulus sindhensis* sp. nov. (Branchiura: Argulidae) from Pakistan. Sindh University Research Journal (Science Series), **43**(2): 199-202.

MAIDL, F., 1912. Beiträge zur Kenntnis des anatomischen Baues der Branchiurengattung *Dolops*. Arbeiten aus dem Zoologischen Instituten der Universität Wien und der Zoologischen Station in Triest, **19**(3): 317-346. [In German.]

MALAVIYA, R. B., 1955. Parasitism of *Ambassis ranga*, H.B. by *Argulus siamensis* subsp. *Peninsularis* Ramakrishna. Current Science, **24**: 275.

MALTA, J. C. O., 1982a. Os argulídeos (Crustacea: Branchiura) da Amazônia Brasileira. Aspectos da ecologia de *Dolops discoidalis* Bouvier, 1899 e *Dolops bidentata* Bouvier, 1899. Acta Amazonica, **12**(3): 521-528. [In Portuguese.]

MALTA, J. C. O., 1982b. Os argulídeos (Crustacea: Branchiura) da Amazônia Brasileira, 2. Aspectos da ecologia de *Dolops geayi* Bouvier, 1897 e *Argulus juparanaensis* Castro, 1950. Acta Amazonica, **12**(4): 701-705. [In Portuguese.]

MALTA, J. C. O., 1983. Os argulídeos (Crustacea: Branchiura) da Amazônia Brasileira 4. Aspectos da ecologia de *Argulus multicolour* Stekhoven, 1937 e *Argulus pestifer* Ringuelet, 1948. Acta Amazonica, **13**(3-4): 489-496. [In Portuguese.]

MALTA, J. C. O., 1984. Os peixes de um lago de Várzea da Amazônia central (Lago Januacá, Rio Solimões) e suas relações com os crustáceos ectoparsitas (Branchiura: Argulidae). Acta Amazonica, **14**(3-4): 355-372. [In Portuguese.]

MALTA, J. C. O. & E. N. SANTOS SILVA, 1986. *Argulus amazonicus* n. sp., crustáceo parasite de peixes da Amazônia Brasileira (Branchiura: Argulidae). Amazoniana, **9**(4): 485-492. [In Portuguese.]

MALTA, J. C. O. & A. VARELLA, 1983. Os argulídeos (Crustacea: Branchiura) da Amazônia Brasileira[3]. Aspectos da ecologia de *Dolops striata* Bouvier, 1899 e *Dolops carvalhoi* Castro, 1949. Acta Amazonica, **13**(2): 299-306. [In Portuguese.]

MALTA, J. C. O. & A. M. B. VARELLA, 2000. *Argulus chicomendesi* sp. n. (Crustacea: Argulidae) parasite de peixes da Amazônia Brasileira. Acta Amazonica, **30**(1): 481-498. [In Portuguese.]

MAMANI, M., C. HAMEL & P. A. VAN DAMME, 2004. Ectoparasites (Crustacea: Branchiura) of *Pseudoplatystoma fasciatum* (surubí) and *P. tigrinum* (chuncuina) in Bolivian white-water floodplains. Ecologia en Bolivia, **39**(2): 9-20.

MARQUES, E., 1978. Copépodes e branquiuros das águas do lago Dilolo (Angola). Garcia de Orta, Serie Zoologica, Lisboa, **7**(1-2): 1-6. [In Portuguese.]

MARTIN, J. W. & G. E. DAVIS, 2001. An updated classification of the recent Crustacea. Natural History Museum of Los Angeles County, 39, 124 pp.

MARTIN, M. F., 1932. On the morphology and classification of *Argulus* (Crustacea). Proceedings of the Zoological Society of London, **103**: 771-806.

MARTÍNEZ, P. R., 1952. *Argulus chilensis*, nov. sp. (Crustacea, Copepoda). Investigaciones Zoologicas Chilenas, **1**(7): 4-9. [In Spanish.]

MBAHINZIREKI, G. B., 1980. Observations on some common parasites of *Bagrus docmac* Forskahl (Pisces: Siluroidea) of Lake Victoria. Hydrobiologia, **75**: 273-280.

MEEHEAN, O. L., 1937. Additional notes on *Argulus trilineatus* (Wilson). Ohio Journal of Science, **37**(5): 288-293.

MEEHEAN, O. L., 1940. A review of the parasitic Crustacea of the genus *Argulus* in the collections of the United States National Museum. Proceedings of the United States National Museum, **88**(3087): 459-522.

MENEZES, J., M. A. RAMOS, T. G. PEREIRA & A. M. DA SILVA, 1990. Rainbow trout culture failure in a small lake as a result of massive parasitosis related to careless fish introductions. Aquaculture, **89**: 123-126.

MERK, T. M., 2016. Effects of antiparasitic treatment for argulosis on innate immune system of a cyprinid fish (fathead minnow, Rafinesque 1820). Dissertation, Faculty of Veterinary Medicine, LMU München, Munich.

MERLA, G., 1961. Zur Bekämpfung der Karpfenlaus (*Argulus*) in der Teichwirtschaft. Deutsche Fischerei-Zeitung, **8**(6): 179-182. [In German.]

MEYER-ROCHOW, V. B., D. AU & E. KESKINEN, 2001. Photoreception in fishlice (Branchiura): the eyes of *Argulus foliaceus* Linne, 1758 and *A. coregoni* Thorell, 1865. Acta Parasitologica, **46**(4): 321-331.

MIKHEEV, V. N., A. V. MIKHEEV, A. F. PASTERNAK & E. T. VALTONEN, 2000. Light-mediated host searching strategies in a fish ectoparasite, *Argulus foliaceus* L. (Crustacea: Branchiura). Parasitology, **120**: 409-416.

MIKHEEV, V. N. & A. F. PASTERNAK, 2010. Parasitic crustaceans influence social relations in fish. Doklady Akademii Nauk, **432**(6): 842-844.

MIKHEEV, V. N., A. F. PASTERNAK & E. T. VALTONEN, 2004. Tuning host specificity during the ontogeny of a fish ectoparasite: behavioural responses to host-induced cues. Parasitology Research, **92**: 220-224.

MIKHEEV, V. N., A. F. PASTERNAK & E. T. VALTONEN, 2007. Host specificity of *Argulus coregoni* (Crustacea: Branchiura) increases at maturation. Parasitology, **134**: 1767-1774.

MIKHEEV, V. N., A. F. PASTERNAK & E. T. VALTONEN, 2015. Behavioural adaptations of argulid parasites (Crustacea: Branchiura) to major challenges in their life cycle. Parasites and Vectors, **8**(1): 1-10.

MIKHEEV, V. N., A. F. PASTERNAK, E. T. VALTONEN & Y. LANKINEN, 2001. Spatial distribution and hatching of overwintered eggs of a fish ectoparasite, *Argulus coregoni* (Crustacea: Branchiura). Diseases of Aquatic Organisms, **46**: 123-128.

MIKHEEV, V. N., E. T. VALTONEN & P. RINTAMÄKI-KINNUNEN, 1998. Host searching in *Argulus foliaceus* L. (Crustacea: Branchiura): the role of vision and selectivity. Parasitology, **116**: 425-430.

MILNE EDWARDS, H., 1834. Histoire naturelle des crustacés, comprenant l'anatomie, la physiologie et la classification de ces animaux. Première tome. (Librairie encyclopédique de Roret, Paris). [In French.]

MILNE EDWARDS, H., 1840. Histoire naturelle des crustacés, comprenant l'anatomie, la physiologie et la classification de ces animaux. Tome troisième. (Librairie encyclopédique de Roret, Paris). [In French.]

MISHRA, T. N. & J. C. CHUBB, 1969. The parasite fauna of the fish of the Shropshire Union Canal, Chesire. Journal of Zoology London, **157**: 213-224.

MØLLER, O. S., 2009. Branchiura (Crustacea) — survey of historical literature and taxonomy. Arthropod Systematics and Phylogeny, **67**(1): 41-55.

MØLLER, O. S., 2012. *Argulus foliaceus*. In: P. T. K. WOO & K. BUCHMANN (eds.), Fish parasites: pathobiology and protection: 327-336. (CAB International, Wallingford).

MØLLER, O. S. & J. OLESEN, 2010. The little-known *Dipteropeltis hirundo* Calman, 1912 (Crustacea, Branchiura): SEM investigations of paratype material in light of recent phylogenetic analyses. Experimental Parasitology, **125**: 30-41.

MØLLER, O. S. & J. OLESEN, 2012. First description of larval stage 1 from a non-African fish parasite *Dolops* (Branchiura). Journal of Crustacean Biology, **32**(2): 231-238.

MØLLER, O. S., J. OLESEN, A. AVENANT-OLDEWAGE, P. F. THOMSEN & H. GLENNER, 2008. First maxillae suction discs in Branchiura (Crustacea): development and evolution in light of the first molecular phylogeny of Branchiura, Pentastomida, and other "Maxillopoda". Arthropod Structure and Development, **37**: 333-346. DOI:10.1016/j.asd.2007.12.002.

MØLLER, O. S., J. OLESEN & D. WALOSZEK, 2007. Swimming and cleaning in the free-swimming phase of *Argulus* larvae (Crustacea, Branchiura)-appendage adaptation and functional morphology. Journal of Morphology, **268**: 1-11. DOI:10.1002/jmor.10491.

MOLNÁR, K. & F. MORAVEC, 1997. *Skrjabillanus cyprinid* n. sp. (Nematoda: Dracunculoidea) from the scales of common carp *Cyprinus carpio* (Pisces) from Hungary. Systematic Parasitology, **38**: 147-151.

MOLNÁR, K. & C. S. SZÈKELY, 1998. Occurrence of skrjabillanid nematodes in fishes of Hungary and in the intermediate host, *Argulus foliaceus* L. Acta Veterinaria Hungarica, **46**(4): 451-463.

MONOD, T. H., 1928. Les Argulidés du Musée du Congo. Revue de Zoologie et de Botanique Africaines, **16**(3): 242-274. [In French.]

MONOD, T. H., 1931. Sur quelques Crustacés aquatiques d'Afrique (Cameroun et Congo). Revue de Zoologie et Botanique Africaines, **21**(1): 1-36. [In French.]

MORAVEC, F., 1978. First record of *Molnaria erythrophthalmi* larvae in the intermediate host in Czechoslovakia. Folia Parasitologica (Praha), **25**: 141-142.

MORAVEC, F., V. VIDAL-MARTÍNEZ & L. AGUIRRE-MACEDO, 1999. Branchiurids (*Argulus*) as intermediate hosts of the daniconematid nematode *Mexiconema cichlasomae*. Folia Parasitologica, **46**: 79.

MOREIRA, C., 1912. Crustacés du Brésil. Mémoires de la Société Zoologique de France, **25**: 145-148. [In French.]

MOREIRA, C., 1913. História Natural, Zoologia, Crustáceos. Comissão de Linhas Telegraphicas Estratégicas de Matto-Grosso ao Amazonas. Papelaria Macedo, **5**: 7-11. [In Portuguese.]

MOREIRA, C., 1915. Les antennes du *Dipteropeltis hirundo* Calman (*Talaus ribeiroi* Moreira) [Crust. Argulidae]. Bulletin de la Société Entomologique de France, **1915**: 120-121. [In French.]

MOUSAVI, H. E., F. BEHTASH, M. ROSTAMI-BASHMAN, S. S. MIRZARGAR, P. SHAYAN & H. RAHMATI-HOLASOO, 2011. Study of *Argulus* spp. infestation rate in goldfish, *Carassius auratus* (Linnaeus, 1758) in Iran. Human and Veterinary Medicine International Journal of the Bioflux Society, **3**(3): 198-204.

MUELLER, J. F., 1936. Notes on some parasitic copepods and a mite, chiefly from Florida fresh water fishes. The American Midland Naturalist, **17**: 807-815.

MÜLLER, O. F., 1785. Entomostraca seu Insecta Testacea, quae in aquis Daniæ et Norvegiæ reperit, descripsit et iconibus illustravit Otho Fridericus Müller. F. W. Thiele, Lipsiae & Havniae Volume: 1-134, index, pls. 1-21. [In Latin.]

NAGASAWA, K., T. AWAKURA & S. URAWA, 1989. A checklist and bibliography of parasites of freshwater fishes of Hokkaido. Scientific Reports of the Hokkaido Fish Hatchery, **44**: 1-49.

NAGASAWA, K. & K. KAWAI, 2008. New host record for *Argulus coregoni* (Crustacea: Branchiura: Argulidae), with discussion on its natural distribution in Japan. Journal of the Graduate School of Biosphere Science, Hiroshima University, **47**: 23-28.

NAGASAWA, K. & S. OHYA, 1996a. *Argulus coregoni* (Crustacea: Branchiura) from Amago salmon *Oncorhynchus masou ishikawai* reared in Central Honshu, Japan. Bulletin of the Fisheries Laboratory of Kinki University, **5**: 83-88. [In Japanese with English summary.]

NAGASAWA, K. & S. OHYA, 1996b. Infection of *Argulus coregoni* (Crustacea: Branchiura) on ayu *Plecoglossus altivelis* reared in central Honshu, Japan. Bulletin of the Fisheries Laboratory of Kinki University, **5**: 89-92.

NANDP, N. C. & S. R. DAS, 1991. Argulosis causing juvenile mortality in some fishes at Kakdwip, west Bengal. Indian Journal of fisheries, **38**(2): 132-133.

NATARAJAN, P., 1982. A new species of *Argulus* Muller (Crustacea: Branchiura), with a note on the distribution of different species of *Argulus* in India. Proceedings — Indian Academy of Sciences, Animal Sciences, **91**(4): 375-380.

NEETHLING, L. A. M. & A. AVENANT-OLDEWAGE, 2015. *Chonopeltis australis* (Crustacea) male reproductive system morphology; sperm transfer and review of reproduction in Branchiura. Journal of Morphology, **276**: 209-218.

NEETHLING, L. A. M., J. C. O. MALTA & A. AVENANT-OLDEWAGE, 2014. Additional morphological information on *Dipteropeltis hirundo* Calman, 1912, and a description of *Dipteropeltis campanaformis* n. sp. (Crustacea: Branchiura) from two characiform benthopelagic fish hosts from two northern rivers of the Brazilian Amazon. Zootaxa, **3755**(2): 179-193.

NETTOVICH, L., 1900. Neue Beiträge zur Kenntniss der Arguliden. Arbeiten aus dem Zoologischen Institut der Universität Wien, **13**(1): 1-32. [In German.]

NOAMAN, V., Y. CHELONGAR & A. H. SHAHMORADI, 2010. The first record of *Argulus foliaceus* (Crustacea: Branchiura) infestation on lionhead goldfish (*Carassius auratus*) in Iran. Iranian Journal of Parasitology, **5**(2): 71-76.

NORTHCOTT, S. J., A. R. LYNDON & A. D. CAMPBELL, 1997. An outbreak of freshwater fish lice, *Argulus foliaceus* L., seriously affecting a Scottish stillwater fishery. Fisheries Management and Ecology, **4**: 73-75.

OAKLEY, T. D., J. M. WOLFE, A. R. LINDGREN & A. K. ZAHAROFF, 2013. Phylotranscriptomics to bring the understudied into the fold: monophyletic Ostracoda, fossil placement, and pancrustacean phylogeny. Molecular Biology and Evolution, **30**(1): 215-233.

OBIEKEZIE, A. I., H. MÖLLER & K. ANDERS, 1988. Diseases of the African estuarine catfish *Chrysichthys nigrodigitatus* (Lacépède) from the Cross River estuary. Journal of Fish Biology, **32**: 207-221.

OKAEME, A. N., A. I. OBIEKEZIE, J. LEHMAN, E. E. ANTAI & C. T. MADU, 1988. Parasites and diseases of cultured fish of Lake Kainji area, Nigeria. Journal of Fish Biology, **32**: 479-481.

ØKLAND, K. A., 1985. Om fiskelus *Argulus* — bygning og levevis, samt registrerte funn I Norge. Fauna, **38**: 53-59. [In Norwegian with English summary.]

ÖKTENER, A., A. H. ALI, A. GUSTINELLI & M. L. FIORAVANTI, 2006. New host records for fish louse, *Argulus foliaceus* L., 1758 (Crustacea, Branchiura) in Turkey. Ittiopatologia, **3**: 161-167.

ÖKTENER, A., J. P. TRÍLLES & I. LEONARDOS, 2007. Five ectoparasites from Turkish fish. Türkiye Parazitoloji Dergisi, **31**(2): 154-157.

OLSON, A. C., 1972. *Argulus melanostictus* and other parasitic crustaceans on the California Grunion, *Leuresthes tenuis* (Osteichthyes: Atherinidae). The Journal of Parasitology, **58**(6): 1201-1204.

OTACHI, E. O., B. SZOSTAKOWSKA, F. JIRSA & C. FELLNER-FRANK, 2015. Parasite communities of the elongate tigerfish *Hydrocynus forskahlii* (Cuvier 1819) and redbelly tilapia *Tilapia zillii* (Gervais 1848) from Lake Turkana, Kenya: influence of host sex and size. Acta Parasitologica, **60**(1): 9-20.

ÖZAN, S. T. & İ. KIR, 2005. An investigation of parasites of goldfish (*Carassius Carassius* L., 1758) in Kovada Lake. Türkiye Parazitoloji Dergisi, **29**(3): 200-203. [In Turkish with English summary.]

ÖZTÜRK, M. O., 2005. An investigation of metazoan parasites of common carp (*Cyprinus carpio* L.) in Lake Eber, Afyon, Turkey. Türkiye Parazitoloji Dergisi, **29**(3): 204-210. [In Turkish with English summary.]

PAIVA CARVALHO, J., 1939. Sôbre dois parasitos do gênero *Dolops* encontrados em peixes de agua dôce. Revista de Industria Animal, **2**(4): 109-116. [In Portuguese.]

PAIVA CARVALHO, J., 1941. Sôbre *Dipteropeltis hirundo* Calman, Crustáceo (Branchiura) parasito de peixes d'agua doce. Boletim da Faculdade de Filosofia, Ciências e Letras, Universidade de Sao Paulo, **22**(5): 265-275. [In Portuguese.]

PAPERNA, I., 1964. Parasitic Crustacea (Copepoda and Branchiura) from inland water fishes of Israel. Israel Journal of Zoology, **13**: 58-68.

PARVEZ, M. M., M. A. B. BHUYAIN, A. M. SHAHABUDDIN, A. R. FARQUE & A. S. SHINE, 2013. Environmentally sustainable control measure of *Argulus* in freshwater ponds in Bangladesh. International Journal of Sustainable Agricultural Technology, **9**(1): 64-70.

PASTERNAK, A., V. N. MIKHEEV & E. T. VALTONEN, 2000. Life history characteristics of *Argulus foliaceus* L. (Crustacea: Branchiura) populations in central Finland. Annales Zoologici Fennici, **37**: 25-35.

PASTERNAK, A., V. N. MIKHEEV & E. T. VALTONEN, 2004a. Growth and development of *Argulus coregoni* (Crustacea: Branchiura) on salmonid and cyprinid hosts. Diseases of Aquatic Organisms, **58**: 203-207.

PASTERNAK, A., V. N. MIKHEEV & E. T. VALTONEN, 2004b. Adaptive significance of the sexual size dimorphism in the fish ectoparasite *Argulus coregoni* (Crustacea: Branchiura). Doklady Biological Sciences, **399**: 477-480.

PEARSE, A. S. M., 1920. The fishes of Lake Valencia, Venezuela. University of Wisconsin Studies in Science, **12**(1): 1-51.

PEARSE, A. S. M., 1924. The parasites of lake fishes. Transactions of the Wisconsin Academy of Sciences, Arts, and Letters, **21**: 161-194.

PEKMEZCI, G. Z., B. YARDIMCI, C. S. BOLUKBAS, Y. E. BEYHAN & S. UMUR, 2011. Mortality due to heavy infestation of *Argulus foliaceus* (Linnaeus, 1758) (Branchiura) in pond-reared carp, *Cyprinus carpio* L., 1758 (Pisces). Crustaceana, **84**: 5-6.

PENCZAK, T., 1972. *Argulus coregoni* Thorell, 1864 (Crustacea, Branchiura) in Poland. Fragmenta Faunistica, **18**(15): 275-282. [In Polish with English summary.]

PEREIRA FONSECA, T., 1939. *Argulus vierai* n. sp., parásito de *Cnesterodon decemmaculatus* (Jenyns). Anales del Museo de Historia Natural de Montevideo, **4**: 3-6. [In Spanish.]

PFEIL-PUTZIEN, C. & C. H. BAATH, 1978. Isolation of the *Rhabdovirus carpio* from carp and carp lice (*Argulus foliaceus*). Berliner und Münchener Tierärztliche Wochenschrift, **91**(22): 445-447. [In German with English summary.]

PFEIL-PUTZIEN, C., 1977. New results in the diagnosis of Spring Viraemia of Carp caused by experimental transmission of *Rhabdovirus carpio* with carp louse (*Argulus foliaceus*). Bulletin — Office International des Epizooties, **87**(5-6): 457.

PIASECKI, W. & A. AVENANT-OLDEWAGE, 2008. Diseases caused by Crustacea. In: J. C. EIRAS, H. SEGNER, T. WAHLI & B. G. KAPOOR (eds.), Fish diseases: 1115-1200. (Science Publishers, Enfield, NH).

PILGRIM, R. L. C., 1967. *Argulus japonicus* Thiele, 1900 (Crustacea: Branchiura) — a new record for New Zealand. New Zealand Journal of Marine and Freshwater Research, **1**: 395-398.

PINEDA, R., S. PÁRAMO & R. DEL RÍO, 1995. A new species of the genus *Argulus* (Crustacea: Branchiura) parasitic on *Astractosteus tropicus* (Pices: Lepisosteidae) from Tabasco, Mexico. Systematic Parasitology, **30**: 199-206.

POLY, W. J., 1997. Host and locality records of the fish ectoparasite, *Argulus* (Branchiura), from Ohio (U.S.A.). Crustaceana, **70**(8): 867-874.

POLY, W. J., 1998a. New state, host, and distribution records of the fish ectoparasite, *Argulus* (Branchiura), from Illinois (U.S.A.). Crustaceana, **71**(1): 1-8.

POLY, W. J., 1998b. *Argulus purpureus* (Risso, 1827), a junior synonym of *Argulus vittatus* (Rafinesque-Schmaltz, 1814) (Branchiura, Arguloida). Crustaceana, **71**(6): 628-632.

POLY, W. J., 2003. *Argulus ambystoma*, a new species parasitic on the salamander *Ambystoma dumerilii* from Mexico (Crustacea: Branchiura: Argulidae). Ohio Journal of Science, **103**(3): 52-61.

POLY, W. J., 2005. *Argulus yucatanus* n. sp. (Crustacea: Branchiura) parasitic on *Cichlasoma urophthalmus* from Yucatan, Mexico. Gulf and Carribean Research, **17**: 1-13.

POLY, W. J., 2008. Global diversity of fishlice (Crustacea: Branchiura: Argulidae) in freshwater. Hydrobiologia, **595**: 209-212.

POULIN, R., 1999. Parasitism and shoal size in juvenile sticklebacks: conflicting selection pressures from different ectoparasites? Ethology, **105**: 959-968.

POULIN, R. & G. J. FITZGERALD, 1987. The potential of parasitism in the structuring of a salt marsh stickleback community. Canadian Journal of Zoology, **65**: 2793-2798.

POULIN, R. & G. J. FITZGERALD, 1988a. Water temperature, vertical distribution, and risk of ectoparasitism in juvenile sticklebacks. Canadian Journal of Zoology, **66**: 2002-2005.

POULIN, R. & G. J. FITZGERALD, 1988b. Risk of parasitism and microhabitat selection in juvenile sticklebacks. Canadian Journal of Zoology, **67**: 14-18.

POULIN, R. & G. J. FITZGERALD, 1989a. A possible explanation for the aggregated distribution of *Argulus canadensis* Wilson, 1916 (Crustacea: Branchiura) on juvenile sticklebacks (Gasterosteidae). Journal of Parasitology, **75**(1): 58-60.

POULIN, R. & G. J. FITZGERALD, 1989b. Male-biased sex ration in *Argulus canadensis* Wilson, 1916 (Crustacea: Branchiura) ectoparasitic on sticklebacks. Canadian Journal of Zoology, **67**: 2078-2080.

POULIN, R. & G. J. FITZGERALD, 1989c. Shoaling as an anti-ectoparasite mechanism in juvenile sticklebacks (*Gasterosteus* spp.). Behavioural Ecology and Sociobiology, **24**: 251-255.

PRICE, R. L. & J. K. BUTTNER, 1980. First reported occurrence of *Argulus mississippiensis* (Crustacea: Branchiura) from Illinois. Transactions of the Illinois State Academy of Science, **72**(3): 8.

PUFFER, H. W. & M. L. BEAL, 1981. Control of parasitic infestations in killifish (*Fundulus parvipinnis*). Laboratory Animal Science, **31**(2): 200-201.

RAFINESQUE-SCHMALTZ, C. S., 1814. Précis des découvertes et travaux somiologiques de C.S. Rafinesque-Schmaltz, entre 1800 et 1814, ou choix raisonné de ses principales découvertes en Zoologie et en Botanique, pour servir d'introduction à ses ouvrages futurs. Palerme Royal Typographie Militaire. [In French.]

RAMAKRISHNA, G., 1951. Notes on the Indian species of the genus *Argulus* Müller (Crustacea, Copepoda) parasitic on fishes. Indian Museum Calcutta Records, **49**(2): 207-215.

RAMDANE, Z. & J.-P. TRILLES, 2012. *Argulus vittatus* (Rafinesque-Schmaltz, 1814) (Crustacea: Branchiura) parasitic on Algerian fishes. Parasitology Research, **110**(4): 1501-1507.

RANGNEKAR, M. P., 1957. Copepod parasites of the families Argulidae, Caligidae, Dichelesthidae, and Lernacopodidae. Journal of the University of Bombay, **26**: 8-20.

REGIER, J. C., J. W. SHULTZ, A. ZWICK, A. HUSSEY, B. BALL & R. WETZER, 2010. Arthropod relationships revealed by phylogenomics analysis of nuclear protein-coding sequences. Nature, **463**(7284): 1079. DOI:10.1038/nature08742.

RILEY, J., A. A. BANAJA & J. L. JAMES, 1978. The phylogenetic relationships of the Pentastomida: the case for their inclusion within the Crustacea. International Journal for Parasitology, **8**: 245-254.

RINGUELET, R., 1943. Revisión de los Argúlidos Argentinos (Crustáces, Branchiua) con el catálogo de las especies Neotropicales. Revista del Museo de La Plata (Nueva Serie), **3**: 43-99. [In Spanish.]

RINGUELET, R., 1948. Argúlidos del Museo de La Plata. Revista de Museo de La Plata (Nueva Serie), **5**: 281-296. [In Spanish.]

RISSO, A., 1826. Histoire naturelle des principals productions de l'Europe méridionale et particulièrement de celles des environs de Nice et des Alpes maritimes. F. G. Levrault, Paris, **5**: 1-403. [In French.]

RIZVI, S. S. H., 1969. Studies on the structure and seasonal incidence of *Argulus foliaceus* (L., 1758) on some freshwater fishes (Branchiura, Argulidae). Crustaceana, **17**(2): 200-206.

ROBERTS, L. S., 1957. Parasites of the carp, *Cyprinus carpio* L. in Lake Taxoma, Oklahoma. The Journal of Parasitology, **43**(1): 54.

ROLAND, C., 1963. Études sur les crustacés Branchioures d'Europe III. Redescription d'*Argulus coregoni* Thorell. Bulletin du Museum National d'Histoire Naturelle, **35**(5): 496-506. [In French.]

ROMANOVSKY, A., 1955. K systematice a rozšíření kapřivců (Argulus) v Československu. Acta Societatis Zoologicae Bohemoslovenicae, **19**(1): 27-43. [In Russian with English summary.]

RUANE, N., T. K. MCCARTHY & P. REILLY, 1995. Antibody response to crustacean ectoparasites in rainbow trout, *Oncorhynchus mykiss* (Walbaum), immunized with *Argulus foliaceus* L. antigen extract. Journal of Fish Diseases, **18**: 529-537.

RUANE, N. M., D. T. NOLAN, J. ROTLLANT, L. TORT, P. H. M. BALM & S. E. WENDELAAR BONGA, 1999. Modulation of the response of rainbow trout (*Oncorhynchus mykiss* Walbaum) to confinement, by an ectoparasitic (*Argulus foliaceus* L.) infestation and cortisol feeding. Fish Physiology and Biochemistry, **20**: 43-51.

RUSHTON-MELLOR, S. K., 1991. *Argulus papuensis* n. sp., a new fish louse (Crustacea: Branchiura) from Papua New Guinea. Systematic Parasitology, **18**: 67-75.

RUSHTON-MELLOR, S. K., 1994a. The genus *Argulus* (Crustacea: Branchiura) in Africa: two new species, *A. fryeri* and *A. gracilis*, the previously undescribed male of *A. brachypeltis* Fryer and the identity of the male described as *A. ambloplites* Wilson. Systematic Parasitology, **28**: 23-31.

RUSHTON-MELLOR, S. K., 1994b. The genus *Argulus* (Crustacea: Branchiura) in Africa: redescription of type-material collected by W.A Cunnington during the Lake Tanganyika expedition in 1913, with notes on *A. giganteus* Lucas and *A. arcassonensis* Cuénot. Systematic Parasitology, **28**: 33-49.

RUSHTON-MELLOR, S. K. & G. A. BOXSHALL, 1994. The development sequence of *Argulus foliaceus* (Crustacea: Branchiura). Journal of Natural History, **28**: 763-785.

SAHA, S. K., A. GUHA & A. BENERJEE, 2011. Feeding apparatus and associated glands in the freshwater fish ectoparasite *Argulus siamensis* Wilson, 1926 (Branchiura). Crustaceana, **84**(10): 1153-1168.

SAHOO, P. K., HEMAPRASANTH, B. KAR, S. K. GARNAYAK & J. MOHANTY, 2012. Mixed infection of *Argulus japonicus* and *Argulus siamensis* (Branchiura, Argulidae) in carps (Pisces, Cyprinidae): loss estimation and a comparative invasive pattern study. Crustaceana, **85**(12-13): 1449-1462.

SAHOO, P. K., B. KAR, A. MOHAPATRA & J. MOHANTY, 2013. *De novo* whole transcriptome analysis of the fish louse, *Argulus siamensis*: first molecular insights into characterization of toll downstream signalling molecules of crustaceans. Experimental Parasitology, **135**: 629-641.

SANDEMAN, J. M. & J. H. C. PIPPY, 1967. Parasites of freshwater fishes Salmonidae and Coregonidae of insular Newfoundland. Journal of the Fisheries Research Board of Canada, **24**(9): 1911-1943.

SAURABH, S., P. K. SAHOO, B. R. MOHANTY, J. MOHANTY, J. K. JENA, S. C. MUKHERJEE & N. SARANGI, 2010. Modulation of the innate immune response of rohu *Labeo rohita* (Hamilton) be experimental freshwater lice *Argulus siamensis* (Wilson) infection. Aquaculture Research, **41**: 326-335.

SCHLÜTER, U., 1978. Observations about host-attacking by the common fish louse *Argulus foliaceus* L. (Crustacea, Branchiura). Zoologischer Anzeiger Jena, **200**(1-2): 85-91. [In German with English summary.]

SCHLÜTER, U., 1979. Über die Temperaturabhängigkeit des Wachstums und Häutungszyklus von *Argulus foliaceus* (L.) (Branchiura). Crustaceana, **37**(1): 100-106. [In German with English summary.]

SCHRAM, T. A., L. IVERSEN, P. A. HEUCH & E. STERUD, 2005. *Argulus* sp. (Crustacea: Branchiura) on cod, *Gadus morhua* from Finnmark, northern Norway. Journal of the Marine Biological Association of the United Kingdom, **85**: 81-86.

SCHUMACHER, R. E., 1952. *Argulus* outbreaks in Minnesota lakes. The Progressive Fish-Culturist, **14**(2): 70.

SCHUURMANS STEKHOVEN, J. H. M., 1937. Crustacea Parasitica. I. Parasitica Copepoda. Résultats Scientifiques des Croissières du Navire Ecole "Mercator". Mémoires du Musée Royal d'Histoire Naturelle de Belgique, **9**(2): 11-24.

SCHUURMANS STEKHOVEN, J. H. M., 1951. Investigaciones sobre argulidos Argentinos. Acta Zoologica Lilloana, **12**: 479-494. [In Spanish.]

SCOTT, T. & A. SCOTT, 1913. The British parasitic Copepoda, Vol. 1: Copepoda parasitic on fishes. (Ray Society, London).

SENG, L. T., 1986. Two ectoparasitic crustaceans belonging to the family Argulidae (Crustacea: Branchiura) in Malaysian freshwater fishes. Malaysian Nature Journal, **39**: 157-164.

SHAFIR, A. & W. H. OLDEWAGE, 1992. Dynamics of a fish ectoparasite population: opportunistic parasitism in *Argulus japonicus* (Branchiura). Crustaceana, **62**(1): 50-64.

SHAFIR, A. & J. G. VAN AS, 1986. Laying, development and hatching of eggs of the fish ectoparasite *Argulus japonicus* (Crustacea: Branchiura). Journal of Zoology: Proceedings of the Zoological Society of London, **210**: 401-414.

SHEN, C. J., 1948. On three new species of fish parasites of the family Argulidae (Crustacea Branchiura). Contributions from the institute of Zoology National Academy of Peiping, **4**(4): 155-166.

SHIMURA, S., 1981. The larval development of *Argulus coregoni* Thorell (Crustacea: Branchiura). Journal of Natural History, **15**: 331-348.

SHIMURA, S., 1983a. Seasonal occurrence, sex ratio and site preference of *Argulus coregoni* Thorell (Crustacea: Branchiura) parasitic on cultured freshwater salmonids in Japan. Parasitology, **86**: 537-552.

SHIMURA, S., 1983b. SEM observation on the mouth tube and preoral sting of *Argulus coregoni* Thorell and *Argulus japonicus* Thiele (Crustacea: Branchiura). Fish Pathology, **18**(3): 151-156.

SHIMURA, S. & M. ASAI, 1984. *Argulus americanus* (Crustacea: Branchiura) parasitic on the bowfin, *Amia calva*, imported from North America. Fish Pathology, **18**(4): 199-203. [In Japanese with English summary.]

SHIMURA, S. & S. EGUSA, 1980. Some ecological notes on the egg deposition of *Argulus coregoni* Thorell (Crustacea, Branchiura). Fish Pathology, **15**(1): 43-47. [In Japanese with English summary.]

SHIMURA, S. & K. INOUE, 1984. Toxic effects of extract from the mouth-parts of *Argulus coregoni* Thorell (Crustacea: Branchiura). Bulletin of the Japanese Society of Scientific Fisheries, **50**(4): 729.

SHIMURA, S., K. INOUE, K. KASAI & M. SAITO, 1983b. Hematological changes of *Oncorhynchus masou* (Salmonidae) caused by the infection of *Argulus coregoni* (Crustacea: Branchiura). Fish Pathology, **18**(3): 157-162. [In Japanese with English summary.]

SHIMURA, S., K. INOUE, M. KUDO & S. EGUSA, 1983a. Studies on effects of parasitism of *Argulus coregoni* (Crustacea: Branchiura) on furunculosis of *Oncorhynchus masou* (Salmonidae). Fish Pathology, **18**(1): 37-40. [In Japanese with English summary.]

SIKAMA, Y., 1938. On a new species of *Argulus* found in a marine fish in Japan. Journal of the Shanghai Science Institute section 3, **4**: 129-134.

SILVA-SOUZA, A. T., V. D. ABDALLAH, R. K. DE AZEVEDO, F. A. DA SILVA & J. L. LUQUE, 2011. Expanded description of *Dolops bidentata* (Bouvier, 1899) (Branchiura: Argulidae) based on specimens collected on *Pygocentrus nattereri* Kner, 1858 (Characiformes) from Poconé Wetland, MT, Brazil. Brazilian Journal of Biology, **71**(1): 145-149.

SINGHAL, R. N., S. JEET & R. W. DAVIES, 1986. Chemotherapy of six ectoparasitic diseases of cultured fish. Aquaculture, **54**: 165-171.

SINGHAL, R. N., S. JEET & R. W. DAVIES, 1990. The effects of argulosis-saprolehniasis on the growth and production of *Cyprinus carpoi*. Hydrobiologia, **202**: 27-31.

SMIT, N. J., L. L. VAN AS & J. G. VAN AS, 2005. Redescription of *Argulus multipocula* Barnard, 1955 (Crustacea: Branchiura) collected on the west coast of South Africa. Systematic Parasitology, **60**: 75-80.

SMITH, S. I., 1873. Catalogue of the marine invertebrate animals of the southern coast of New England, and adjacent waters in: Invertebrate animals of Vineyard Sound and adjacent waters. United States Commissioner of Fisheries. Part 1. Report on the condition of the sea fisheries of the South Coast of New England in 1871 and 1872: 573-578.

STAMMER, J., 1959. Beiträge zur Morphologie, Biologie und Bekämpfung der Karpfenläuse. Zeitschrift für Parasitenkunde, **19**: 135-208. [In German.]

STUHLMANN, F., 1891. Zur Kenntniss der Fauna central-afrikanischer Seen. II. Ueber eine neue Art der Arguliden-Gattung. Zoologische Jahrbücher abtheilung für systematic, geographie und biologie der thiere, **6**: 152-154. [In German.]

SUNDARA BAI, A., D. SEENAPPA & K. V. DEVERAJ, 1988. Oviposition and sex ratio of *Argulus siamensis* var. *siamensisi* and *Argulus siamensis* var. *hessarghattaris* (Crustacea: Branchiura) parasitic on freshwater fishes. Current Science, **57**(12): 685-686.

SUTHERLAND, D. R. & D. D. WITTROCK, 1986. Surface topography of the branchiuran *Argulus appendiculosus* Wilson, 1907 as revealed by scanning electron microscopy. Zeitschrift für Parasitenkunde, **72**: 405-415.

SWANEPOEL, J. H. & A. AVENANT-OLDEWAGE, 1992. Comments on the morphology of the pre-oral spine in *Argulus* (Crustacea: Branchiura). Journal of Morphology, **212**: 155-162.

SWANEPOEL, J. H. & A. AVENANT-OLDEWAGE, 1993. Functional morphology of the foregut of *Chonopeltis australis* Boxshall (Branchiura). Journal of Crustacean Biology, **13**(4): 656-666.

SZALAI, A. J. & T. A. DICK, 1991. Evaluation of gill nets, fyke nets, and mark-recapture methods to estimate the number of Hirudinea and Crustacea on fish. The Journal of Parasitology, **77**(6): 914-922.

TAKEMOTO, R. M., G. C. PAVENELLI, M. A. P. LIZAMA, A. C. F. LACERDA, F. H. YAMADA, L. H. A. MOREIRA, T. L. CESCHINI & S. BELLAY, 2009. Diversity of parasites of fish from the Upper Paraná River floodplain, Brazil. Brazilian Journal of Biology, **69**(2, Suppl.): 691-705.

TAM, Q. & A. AVENANT-OLDEWAGE, 2006. The digestive system of larval *Argulus japonicus* (Branchiura). Journal of Crustacean Biology, **26**(4): 447-454.

TAM, Q. & A. AVENANT-OLDEWAGE, 2009a. The ultrastructure of the digestive cells of *Argulus japonicus* Thiele, 1900 (Crustacea: Branchiura). Arthropod Structure & Development, **38**: 45-53.

TAM, Q. & A. AVENANT-OLDEWAGE, 2009b. The effect of starvation on the ultrastructure of the digestive cells of *Dolops ranarum* (Stuhlmann, 1891) (Crustacea: Branchiura). Arthropod Structure & Development, **38**: 391-399.

TAM, Q., A. AVENANT-OLDEWAGE & E. H. WILLIAMS, 2005. An ultrastructural investigation of *Argulus personatus* Cunnington, 1913 (Crustacea: Branchiura) from Lake Tanganyika, northern Zambia. African Zoology, **40**(2): 301-308.

TAVARES-DIAS, M., F. R. DE MORAES, E. M. ONAKA & P. C. B. REZENDE, 2007. Changes in blood parameters of hybrid tambacu fish parasitized by *Dolops carvalhoi* (Crustacea, Branchiura), a fish louse. Veterinarski Archiv, **77**(4): 355-363.

TAYLOR, N. G. H., C. SOMMERVILLE & R. WOOTTEN, 2006. The epidemiology of *Argulus* spp. (Crustacea: Branchiura) infections in Stillwater trout fisheries. Journal of fish Diseases, **29**: 193-200.

TAYLOR, N. G. H., R. WOOTTEN & C. SOMMERVILLE, 2009a. The influence of risk factors on the abundance, egg laying habits and impact of *Argulus foliaceus* in Stillwater trout fisheries. Journal of Fish Diseases, **32**: 509-519.

TAYLOR, N. G. H., R. WOOTTEN & C. SOMMERVILLE, 2009b. Using length-frequency data to elucidate the population dynamics of *Argulus foliaceus* (Crustaca: Branchiura). Parasitology, **136**: 1023-1032.

TEKIN ÖZAN, S. & I. KIR, 2005. An investigation of parasites of goldfish (*Carassius carassius* L., 1758) in Kovada Lake. Türkiye Parasitoloji Dergisi, **29**(3): 200-203. [In Turkish with English summary.]

THATCHER, V. E., 1991. Amazon fish parasites. Amazoniana: limnologia et oecologia regionalis systemae fluminis Amazonas, **11**(3-4): 263-572.

THATCHER, V. E., 2006. Branchiura. In: J. ADIS, J. R. ARIAS, G. RUEDA-DELGADO & K. M. WANTZEN (eds.), Aquatic biodiversity in Latin America, Vol. 1: Amazon fish parasites (2nd ed.): 390-415. (Pensoft, Sofia).

THIELE, J., 1900. Diagnosen neuer Arguliden-Arten. Zoologischer Anzeiger, **22**: 46-48. [In German.]

THIELE, J., 1904. Beiträge zur Morphologie der Arguliden. Mitteilungen aus dem Zoologischen Museum in Berlin 2, **4**(1904): 1-51. [In German.]

THILAKARATNE, I. D. S. I. P., G. RAJAPAKSHA, A. HEWAKOPARA, R. P. V. J. RAJAPAKSE & A. C. M. FAIZAL, 2003. Parasitic infections in freshwater ornamental fish in Sri Lanka. Diseases of Aquatic Organisms, **54**: 157-162.

THOMAS, M. M., 1961. Observation on the habits and post-embryonic development of a parasitic branchiuran *Argulus puthenveliensis* Ramakrishna. Journal of the Marine Biology Association India, **3**(1-2): 75-86.

THOMAS, M. M. & M. DEVERAJ, 1975. Two new species of *Argulus* Muller (Crustacea: Branchiura) from River Cauvery with a key to Indian species. Indian Journal of fisheries, **22**(1-2): 215-220.

THOMSEN, R., 1925. "*Argulus violaceus*" nov. spec. Cangrejo parasite del bagre. Physis, **8**: 185-198. [In Spanish.]

THOMSEN, R., 1942. Notas criticas acerca de dos Argúlidos (Branchiura) del Brasil. Anais da Academia Brasileira de Ciências, **14**: 37-45. [In Spanish.]

THORELL, M. T., 1864. Om tvenne Europeiske Argulider, jemte anmarkninger am Argulider nas morfolog och systematiska stallning, samt en ofversigt af de for navarande kanda arterna of denna familj. Annals and Magazine of Natural History, **3**(18): 149-169. [In Swedish.]

THURSTON, J. P., 1970. The incidence of Monogenea and parasitic Crustacea on the gills of fish in Uganda. Revue de Zoologie et de Botanique Africaines, **82**(1-2): 111-130.

TIDD, W. M., 1931. A list of parasitic copepods and their fish hosts from Lake Erie. Ohio Journal of Science, **31**(6): 453-454.

TIKHOMIROVA, V. A., 1970. Life cycle of the nematode *Skrjabillanus scardinii* Molnar, 1965. Doklady Akademii Nauk SSSR, **195**(2): 510-511. [In Russian with English Summary.]

TIKHOMIROVA, V. A., 1980. On nematodes of the family Skrjabillanidae (Nematoda: Camallanata). Parazitologiia, **14**(3): 258-262. [In Russian with English Summary.]

TOKIOKA, T., 1936a. Preliminary report on Argulidae found in Japan. Annotationes Zoologicae Japonenses, **15**(3): 334-343.

TOKIOKA, T., 1936b. Larval development and metamorphosis of *Argulus japonicus*. Memoirs of the College of Science, Kyoto Imperial University, Series B, **12**(1): 93-114.

TOKIOKA, T., 1939. *Argulus* of Manchukuo. Annotationes Zoologicae Japonenses, **18**(1): 42-45.

UMA, A., G. REBECCA & K. SARAVANABAVA, 2012. Differential expression of toll-like receptors (TLRs) in gold fish, *Carassius auratus* infested with fresh water lice of *Argulus* sp. International Journal of Pharmacy and Biological Sciences, **3**(4): 652-658.

UZUNAY, E. & E. SOYLU, 2006. Metazoan parasites of carp (*Cyprinus carpio* Linnaeus, 1758) and vimba (*Vimba vimba* Linnaeus, 1758) in the Sapanca Lake. Türkiye Parasitoloji Dergisi, **30**(2): 141-150. [In Turkish with English summary.]

VALLADÃO, G. M. R., S. U. GALLANI & F. PILARSKI, 2015. Phytotherapy as an alternative for treating fish disease. Journal of Veterinary Pharmacology and Therapeutics, **38**(5): 417-428.

VAN AS, J. G., 1986. A new species of *Chonopeltis* (Crustacea: Branchiura) from the Limpopo system, southern Africa. South African Journal of Zoology, **21**(4): 348-351.

VAN AS, J. G., 1992. A new species of *Chonopeltis* (Crustacea: Branchiura) from the Zambesi River system. Systematic Parasitology, **22**: 221-229.

VAN AS, J. G. & L. L. VAN AS, 1999c. *Chonopeltis liversedgei* sp. n. (Crustacea: Branchiura), parasite of the western bottlenose *Mormyrus lacerda* (Mormyridae) from the Okavango delta, Botswana. Folia Parasitologica, **46**: 319-325.

VAN AS, J. G. & L. L. VAN AS, 2001. *Argulus izintwala* n. sp. (Crustacea: Branchiura) from Lake St. Lucia, South Africa. Systematic Parasitology, **48**: 75-79.

VAN AS, J. G., J. P. VAN NIEKERK & P. A. S. OLIVIER, 1999. Description of the previously unknown male of *Argulus kosus* Avenant-Oldewage, 1994 (Crustacea: Branchiura). Systematic Parasitology, **43**: 75-80.

VAN AS, L. L. & J. G. VAN AS, 1993. First record of *Chonopeltis inermis* Thiele, 1900 (Crustacea: Branchiura) in the Limpopo River system with notes on its morphology. Systematic Parasitology, **24**: 229-236.

VAN AS, L. L. & J. G. VAN AS, 1996. A new species of *Chonopeltis* (Crustacea: Branchiura) from the southern Rift Valley, with notes on larval development. Systematic Parasitology, **35**: 69-77.

VAN AS, L. L. & J. G. VAN AS, 1999a. Aspects of the morphology and a review of the taxonomic status of three species of the genus *Chonopeltis* (Crustacea: Branchiura) from the Orange-Vaal and South West Cape River systems, South Africa. Folia Parasitologica, **46**: 221-228.

VAN AS, L. L. & J. G. VAN AS, 1999b. Male copulatory structures as a taxonomic tool in the genus *Chonopeltis* Thiele, 1900. Microscopy Society of Southern Africa Proceedings, **29**: 78.

VAN AS, L. L. & J. G. VAN AS, 2015. Branchiuran parasites (Crustacea: Branchiura) from fishes in the Okavango (Botswana) and Zambezi (Namibia) systems. African Journal of Aquatic Science, **40**(1): 9-20.

VAN BENEDEN, P. J., 1891. Un Argule nouveau des côtes d'Afrique. Bulletin de l'Académie Royal de Belgique, 3e série, **22**(11): 369-378. [In French.]

VAN DEN BOSCH DE AGUILAR, P., 1972. Le systéme neurosécréteur de l'*Argulus foliaceus* (L.) (Branchiura). Crustaceana, **23**: 123-132. [In French.]

VAN DER SALM, A. L., D. T. NOLAN, F. A. T. SPANINGS & S. E. WENDELAAR BONGA, 2000. Effects of infection with the ectoparasite *Argulus japonicus* (Thiele) and administration of cortisol on cellular proliferation and apoptosis in the epidermis of common carp, *Cyprinus carpio* L., skin. Journal of Fish Diseases, **23**: 173-184.

VAN KAMPEN, P. N., 1909. Über *Argulus belones* n. sp. und *A. indicus* M. Weber aus dem Indischen Archipel. Zoologischer Anzeiger, **34**(13/14): 443-447. [In German.]

VAN NIEKERK, J. P. & D. J. KOK, 1989. *Chonopeltis australis* (Branchiura): structural, developmental and functional aspects of the trophic appendages. Crustaceana, **57**(1): 51-56.

VONDRKA, K., 1972. Efficacy of antiparasitic baths against fish lice, employing variously hard water. Acta Veterinaria Brno, **41**: 113-116.

WADEH, H., J. W. YANG & G. Q. LI, 2008. Ultrastructure of *Argulus japonicus* Thiele, 1900 (Crustacea: Branchiura) collected from Guangdong, China. Parasitology Research, **102**: 765-770.

WAGLER, E., 1935. Die deutschen Karpfenläuse. Zoologischer Anzeiger, **110**(1/2): 1-10. [In German.]

WALKER, P. D., G. FLIK & S. E. WENDELAAR BONGA, 2004. The biology of parasites from the genus *Argulus* and a review of the interactions with its host. In: G. F. WIEGERTJES & G. FLIK (eds.), Host-parasite interactions: 107-129. (BIOS Scientific Publishers, Abingdon).

WALKER, P. D., J. E. HARRIS, G. VAN DER VELDE & S. E. WENDELAAR BONGA, 2008a. Differential host utilisation by different life history stages of the fish ectoparasite *Argulus foliaceus* (Crustacea: Branchiura). Folia Parasitologica, **55**: 141-149.

WALKER, P. D., J. E. HARRIS, G. VAN DER VELDE & S. E. WENDELAAR BONGA, 2008b. Effect of host weight on the distribution of *Argulus foliaceus* (L.) (Crustacea, Branchiura) within a fish community. Acta Parasitologica, **53**(2): 165-172.

WALKER, P. D., I. J. RUSSON, R. DUIJF, G. VAN DER VELDE & S. E. WENDELAAR BONGA, 2011a. The off-host survival and viability of a native and non-native fish louse (*Argulus*, Crustacea: Branchiura). Current Zoology, **57**(6): 828-835.

WALKER, P. D., I. J. RUSSON, C. HAOND, G. VAN DER VELDE & S. E. WENDELAAR BONGA, 2011b. Feeding in adult *Argulus japonicus* Thiele, 1900 (Maxillopoda, Branchiura), an ectoparasite on fish. Crustaceana, **84**(3): 307-318.

WALTER, T. C., 2008. *Gyropeltis* Heller, 1857. In: T. C. WALTER & G. BOXSHALL (2015). World of Copepods database. World Register of Marine Species, available online at http://www.marinespecies.org/aphia.php?p=taxdetails&id=347204 (accessed 28 June 2016).

WALTER, T. C., 2015a. *Dolops* Audouin, 1837. In: T. C. WALTER & G. BOXSHALL (2015). World of Copepods database. World Register of Marine Species, available online at http://www.marinespecies.org/aphia.php?p=taxdetails&id=347079 (accessed 28 June 2016).

WALTER, T. C., 2015b. *Chonopeltis* Thiele, 1900. In: T. C. WALTER & G. BOXSHALL (2015). World of Copepods database. World Register of Marine Species, available online at http://www.marinespecies.org/aphia.php?p=taxdetails&id=347973 (accessed 28 June 2016).

WANG, K.-N., 1958. Preliminary studies on four species of *Argulus* parasitic on freshwater fishes taken from the area between Nanking and Shanghai, with notes on the early larval development of *Argulus chinensis*. Acta Zoologica Sinica, **10**: 322-340. [In Chinese with English summary.]

WANG, K.-N., 1960. Two new species of *Argulus* from freshwater fishes in China. Acta Zoologica Sinica, **12**: 242-247. [In Chinese with English summary.]

WATSON, R. & A. AVENANT-OLDEWAGE, 1996. Damage caused by the attachment of *Argulus japonicus* to its host. Miscroscopy Society of Southern Africa Proceedings, **26**: 126.

WATSON, R. A. & T. A. DICK, 1979. Metazoan parasites of whitefish *Coregonus clupeaformis* (Mitchill) and cisco *C. artedii* LeSueur from Southern Indian Lake, Manitoba. Journal of Fish Biology, **15**: 579-587.

WEBB, A. C., 2008. Spatial and temporal influences on population dynamics of a branchiuran ectoparasite, *Argulus* sp. A, in fresh waters of tropical northern Queensland Australia. Crustaceana, **81**: 1055-1067.

WEBER, M., 1892. Die Süsswasser Crustaceen des Indischen Archipels, nebst Bemerkungen über die Süsswasser Fauna im Allgemeinen. Zoologische Ergebnisse einer Reise in Niederländisch Ost Indien, **2**: 528-571. [In German.]

WEIBEZAHN, F. H. & T. COBO, 1964. Seis argulidos (Crustacea, Branchiura) parasitos de peces dulce-acuicolas en Venezuela, con descripcion de una nueva especie del genero *Argulus*. Acta Biologica Venezuelica, **4**(2): 119-144. [In Spanish.]

WILSON, C. B., 1902. North American parasitic copepods of the family Argulidae, with a bibliography of the group and a systematic review of all known species. Proceedings of the United States National Museum, **25**(1302): 635-742.

WILSON, C. B., 1904. A new species of *Argulus* with a more complete account of two species already described. Proceedings of the United States National Museum, **27**(1368): 627-655.

WILSON, C. B., 1907. Additional notes on the development of the Argulidae with descriptions of a new species. Proceedings of the United States National Museum, **32**(1531): 411-424.

WILSON, C. B., 1909. North American parasitic copepods: a list of those found upon the fishes of the Pacific coast, with a description of new genera and species. Proceedings of the United States National Museum, **35**(1652): 432-481.

WILSON, C. B., 1912a. Parasitic copepods from Nainamo, British Columbia, including eight species new to science. Contributions to Canadian Biology, being studies from the Marine Biological Stations of Canada, 1906-1910: 85-101.

WILSON, C. B., 1912b. Descriptions of new species of parasitic copepods in the collections of the United States National Museum. Proceedings of the United States National Museum, **42**(1900): 233-243.

WILSON, C. B., 1916. Copepod parasites of freshwater fishes and their economic relations to mussel glochidia. Bulletin of the Bureau of Fisheries, **34**(824): 1-76.

WILSON, C. B., 1920a. Argulidae from the Shubenacadie River, Nova Scotia. The Canadian field-naturalist, **34**(1): 149-151.

WILSON, C. B., 1920b. Parasitic copepods from the Congo Basin. Bulletin of the American Museum of Natural History, **43**: 1-8.

WILSON, C. B., 1922. Parasitic copepods from Japan, including five new species. Arkiv för Zoologi, Stockholm, **14**(10): 1-17.

WILSON, C. B., 1923. New species of parasitic copepods from southern Africa. Meddelanden fran Göteborgs Musei Zoologiska Avdelning, **19**: 1-11.

WILSON, C. B., 1924. New North American parasitic copepods, new hosts, and notes on copepod nomenclature. Proceedings of the United States National Museum, **64**(2507/17): 1-22.

WILSON, C. B., 1926. A new parasitic copepod from Siam. Journal of the Siam Society, Natural History Supplement, **6**(4): 361-363.

WILSON, C. B., 1927. A copepod (*Argulus indicus*) parasitic on the fighting-fish in Siam. The Journal of the Siam Society, Natural History Supplement, **7**(1): 1-3.

WILSON, C. B., 1932. The copepods of the Woods Hole Region, Massachusetts. United States National Museum, Bulletin, **158**: 1-635.

WILSON, C. B., 1935a. Parasitic copepods from the Dry Tortugas. Papers from the Tortugas Laboratory, **29**(7): 329-347.

WILSON, C. B., 1935b. Parasitic copepods from the Pacific coast. American Midland Naturalist, **16**: 776-797.

WILSON, C. B., 1936a. Copepods from the cenotes and caves of the Yucatan Peninsula, with notes on Cladocerans. Publications of the Carnegie Institution of Washington, **457**: 77-88.

WILSON, C. B., 1936b. Two new parasitic copepods from Cuban fish. Memorias de la Sociedad Cubana de Historia Natural, **10**(2): 107-112.

WILSON, C. B., 1944. Parasitic copepods in the United States National Museum. Proceedings of the United States National Museum, **94**(3177): 529-582.

WINGSTRAND, K. G., 1972. Comparative spermatology of a pentastomid, *Raillietiella hemidactyli*, and a branchiuran crustacean, *Argulus foliaceus*, with a discussion of pentastomid relationships. Det Kongelige Danske Videnskabernes Selskab, Biologiske Skrifter, **19**: 635-762.

WOLFE, B. A., C. A. HARMS, J. D. GROVES & M. R. LOOMIS, 2001. Treatment of *Argulus* sp. infestation of river frogs. The American Association for Laboratory Animal Science, **40**(6): 35-36.

YAMAGUTI, S., 1937. On two species of *Argulus* from Japan. Papers on Helminthology, All Union Lenin Academy of Agricultural Sciences, Moscow, pp. 784.

YAMAGUTI, S., 1963. Parasitic Copepoda and Branchiura of fishes: 319-389. (Interscience Publishers, London).

YAMAGUTI, S. & T. YAMASU, 1959. On two species of *Argulus* (Branchiura, Crustacea) from Japanese fishes. Biological Journal of Okayama University, **5**(3-4): 167-175.

YEATMAN, H. C., 1965. Redescription of the freshwater branchiuran crustacean, *Argulus diversus* Wilson, with a comparison of related species. The Journal of Parasitology, **51**(1): 100-107.

YILDIZ, K. & A. KUMANTAS, 2002. *Argulus foliaceus* infection in a goldfish (*Carassius auratus*). Israel Journal of veterinary medicine, **57**(2): 18-120.

YOSHIZAWA, K. & S. NOGAMI, 2008. The first report of phototaxis of fish ectoparasite, *Argulus japonicus*. Research in Veterinary Science, **85**: 128-130.

ZAĆWILICHOWSKA, K., 1948. The nervous system of the carp-louse *Argulus foliaceus* L. Bulletin International de l'Académie Polonaise des Sciences et des Lettres, Classe des Sciences Mathematiques et Naturelles, Serie B: Sciences Naturelles (II), **1**: 117-128.

ZAĆWILICHOWSKI, J., 1935. Über die Innervation der Haftapparate der Karpfenlaus *Argulus foliaceus* L. (Branchiura). Bulletin International de l'Académie Polonaise des Sciences et des Lettres, Classe des Sciences Mathématiques et Naturelles, Serie B: Sciences Naturelles (II), **1**: 145-162. [Cracovie.]

ZADDACH, E. G., 1844. Synopseos Crustaceorum Prussicorum Prodromus. Dissertation Zoologica, Regiomonti (Habilitationschrift). VI + 39.

ZENKER, W., 1854. System der Crustaceen. Archiv für Naturgeschichte, **20**(1): 108-117.

ZHILIUKAS, V. J. & E. J. RAUCKIS, 1982. *Argulus foliaceus* in young of Coregonidae reared in cages. Parazitologiya, **16**(2): 117-121. [In Russian with English summary.]

ZRZAVÝ, J., 2001. The interrelationships of metazoan parasites: a review of phylum-and higher-level hypotheses from recent morphological and molecular phylogenetic analyses. Folia Parasitologia (Praha), **48**: 81-103.

First received 25 November 2015.
Final version accepted 6 September 2016.

TAXONOMIC INDEX

HOST INDEX